Problems in Unification and Supergravity

(La Jolla Institute 1983)

DEDICATION

These proceedings are dedicated by his friends and colleagues to

MARVIN GOLDBERGER

in celebration of his 60th birthday.

AIP Conference Proceedings
Series Editor: Hugh C. Wolfe
No. 116

Problems in Unification and Supergravity

(La Jolla Institute 1983)

Edited by
Glennys Farrar and Frank Henyey

American Institute of Physics
New York 1984

L.C. Catalog Card No. 84-71246
ISBN 0-88318-315-3
DOE CONF- 830139

v

Marvin Goldberger

DEDICATION

This volume on "Problems in Unification and Supergravity," dedicated to Marvin Goldberger in celebration of his 60th birthday, is the written record of a workshop held in La Jolla early in 1983. Characteristically ahead of his time, Goldberger had actually reached his 60th year on October 22, 1982, a few months in advance of the conference. The published Proceedings, herein, inevitably comes last in this sequence of events.

Goldberger belonged to that remarkable band of graduate students who accumulated, and scintillated, at the University of Chicago just after World War II. During the war itself he had worked on the Manhattan Project, armed with a B.S. degree from the Carnegie Institute of Technology. His first published paper stemmed from that work and foreshadowed the Goldberger to come: written with Frederick Seitz, it was a paper on *scattering* phenomena (of neutrons on crystals). For his Ph.D. dissertation he undertook one of the earliest applications of the Monte Carlo technique (as we now call it), which had recently been developed by Ulam and von Neumann. At Fermi's suggestion he made a theoretical analysis of scattering of high energy neutrons off heavy nuclei. In the account published in the Physical Review, he tells the reader that he followed the history of 100 incident neutrons, by hand computation. Beyond this, he says with youthful awe, "the problem should be handled by a machine."

Subsequently, in the early 1950's, he produced a series of increasingly fundamental, and enormously influential papers on the formal theory of scattering and on the burgeoning new field of causality and dispersion relations. Then, starting later in the decade, a wealth of applications of the new analyticity insights began to issue from his pen. By 1964, Goldberger and his collaborator Kenneth Watson saw through to publication their great and comprehensive treatise *Collision Theory*. This classic has almost everything, as Goldberger will readily acknowledge: "Its in the BOOK, go look it up" he will often say when asked a question he can't immediately answer (there are few such questions, however).

He received his Ph.D. in 1948, sojourned briefly at the Berkeley Radiation Lab and at MIT, then returned to Chicago in 1950 as a member of the faculty. In 1957, he translated to Princeton, for a long and productive stay as teacher, colleague, leader, and from 1970 to 1976, chairman of the physics department. The pursuit and exploitation of "analyticity physics" intensified during his early years at Princeton, broadening, in particular, into the area of the weak interactions. I was granted an early taste of the Goldberger style via a joint piece of research we'd done which led us to an unexpected relation connecting weak and strong interaction parameters. Putting in the experimental numbers, we found what I thought to be spectacular agreement, to within ~ 10 percent. Goldberger, however, was irritated by this residual discrepancy. "The experimental results are wrong," he breezily announced.

While advancing the frontiers of physics through his personal approach, and as the guru of generations of students, postdocs and faculty colleagues at Princeton, Goldberger became increasingly active also on the wider stage of national security affairs, disarmament, international scientific cooperation, and the like. He was a member of the President's Science Advisory Committee (1965-69), chairman of the Federation of American Scientists (1972-73), advisor to various government agencies, and much else. In 1978, he was called to the presidency of the California Institute of Technology - a very large stage indeed.

Murph is universally admired for his personal warmth and loyalty, and for his sense of style in all things. Let it be said, however, that beneath the light-hearted and effortless, there is in him a fierce dedication to high standards in science, and a fierce love of physics. Happy birthday, dear Murph.

Sam Treiman
Princeton University

PREFACE

The understanding of properties of quantum field theory even at energies in excess of those obtained by accelerators has been very productive. Demanding renormalizability and asymptotic freedom leads to gauge theories, which can be extended to unified gauge theories of fundamental physical forces excluding gravitation. These very high energy (short distance) aspects have important consequences at normal energies. The high energy behavior has also been productive in understanding the structure of the universe--extremely high energies occurred immediately following the big bang. It is expected that the inclusion of gravity into these considerations will have equally, or even more, profound implications, both in cosmology and in constraining the possible structure of nature as observed at normal energies.

Theoretical particle physics is at an unusual historical point. The "standard model" with an SU(3) x SU(2) x U(1) gauge group is not in conflict with any well established experimental facts and appears to describe, along with general relativity for gravitation, all presently known phenomena. Nonetheless, we regard it as inadequate for several reasons. We would like to understand why these particular groups are selected, why the fundamental fields are the particular ones which occur, and how the patterns of spontaneous symmetry breaking occur. The tremendous theoretical success of the quantum gauge theories serves to emphasize the lack of an acceptable quantum theory of gravity.

These questions formed the focus of a La Jolla Institute workshop entitled "Unification and Supergravity," held at the Institute January 13-16, 1983. Problems in unification of the various gauge groups, quantum gravity, supersummetry and supergravity, compact dimensions of space-time, and conditions at the beginning of the universe were discussed.

Financial support for the workshop was provided by the La Jolla Institute. Support for these proceedings was provided by the Department of Energy and by the Institute.

Glennys Farrar
Department of Physics
Rutgers University
P.O. Box 849
Piscataway, New Jersey 08854

Frank Henyey
Center for Studies of
Nonlinear Dynamics
La Jolla Institute
Suite 2150
8950 Villa La Jolla Drive
La Jolla, California 92037

LIST OF PARTICIPANTS

ORGANIZING COMMITTEE

Prof. Glennys Farrar	Rutgers University
Dr. Frank Henyey	La Jolla Institute
Prof. Murray Gell-Mann	California Institute of Technology
Prof. Gordon Kane	University of Michigan
Prof. Norman Kroll	University of California, San Diego
Prof. Steven Weinberg	University of Texas

INVITED PARTICIPANTS

Prof. T. Appelquist	Yale University
Prof. Richard Arnowitt	Northeastern University
Prof. Curtis Callan	Princeton University
Dr. Michael Dine	Institute for Advanced Study
Prof. John Ellis	Stanford Linear Accelerator Center
Dr. Sergio Ferrara	CERN
Dr. Venkatesh Ganapathi	University of California, San Diego
Prof. Howard Georgi	Harvard University
Dr. Steven Gottlieb	University of California, San Diego
Dr. John Hagelin	Stanford Linear Accelerator Center
Dr. Lawrence Hall	Lawrence Berkeley Laboratory
Prof. Luis Ibanez	University of Madrid
Prof. Stanley Mandelstam	University of California, Berkeley
Dr. Chiara Nappi	Institute for Advanced Study
Dr. Hans Nilles	CERN
Prof. David Olive	University of Virginia
Dr. Burt Ovrut	Rockefeller University
Prof. Malcolm Perry	Princeton University
Prof. Michael Peskin	Stanford Linear Accelerator Center
Dr. J. Polchinski	Harvard University
Prof. Stuart Raby	University of Michigan
Prof. Jonathan Rosner	University of Chicago
Dr. John Schwarz	California Institute of Technology
Prof. Alberto Sirlin	New York University
Prof. Paul Steinhardt	University of Pennsylvania
Dr. Kellogg Stelle	Imperial College
Prof. Leonard Susskind	Stanford University
Dr. Thomas Weiler	University of California, San Diego
Dr. Peter West	Kings College
Prof. E. Witten	Princeton University

TABLE OF CONTENTS

DIMENSIONAL REDUCTION IN QUANTUM GRAVITY*†

THOMAS APPELQUIST
J.W. GIBBS LABORATORY
YALE UNIVERSITY
NEW HAVEN, CT 06511

ABSTRACT

The phenomenon of dimensional reduction in quantum theories of gravity is described. The five-dimensional Kaluza-Klein model is studied and the one-loop effective potential, as a function of the five-five component of the metric, is computed. For values of this field such that the distance around the fifth dimension is larger than the Planck length, the loop expansion for the effective potential is reliable. The form of the potential then implies the existence of a force tending to make the fifth dimension contract to a size on the order of the Planck length. This result can be interpreted as a gravitational version of the Casimir effect in quantum electrodynamics.

†Research supported in part by the U.S. Department of Energy under Contract No. DE-AC02-76ER03075.

One of the most interesting and attractive ways of unifying gauge theories and gravitation is also one of the oldest - the higher dimensional theory of Kaluza and Klein[1], which, since its introduction sixty years ago, has been developed and generalized by many authors[2].

The usual procedure is to consider a field theory, most simply pure general relativity, in D dimensions, D>4, and then to make an Ansatz (such as, for example, that nothing depends on the extra D-4 coordinates) in order to reduce the theory to an effective four-dimensional one. The question arises, however, whether the extra dimensions serve only as an intermediate device for deriving the four-dimensional theory[3], or whether they really exist in the sense that the four-dimensional theory is to be regarded as an approximation to the full D-dimensional dynamics. Only in the latter case is the Kaluza-Klein theory truly a unified one.

Although it is possible to apply dimensional reduction to theories which do not include gravity[4], it would seem that if one wants to understand the dynamics of dimensional reduction, one should begin with a theory that has within it the possibility of modifying the geometry of space-time. In this talk I will focus on the simplest prototype for such a theory, the original five-dimensional model considered by Kaluza and Klein.

If the extra dimension, or dimensions, really exist, one must explain how it is that they are not seen. The usual answer, that they are extremely small, not more than a few orders of magnitude bigger than the Planck length $(\hbar G/c^3)^{\frac{1}{2}} = 1.6\times10^{-33}$ cm, begs the question since no mechanism for this smallness has, till now, been suggested. Given the smallness of these dimensions, however, one's intuition is that physics should be insensitive to them until distance scales of order the Planck length are being probed.

This is, however, not strictly true. The degrees of freedom which have been frozen out by the process of dimensional reduction can still have an impact on low-energy physics because of their appearance as virtual particles in quantum loops. This effect is well-known, for example, in the case of gauge theories at finite temperature[5]. These, too, can be viewed as dimensionally reduced, with a periodicity in the time coordinate inversely proportional to the temperature. In the phenomenon of Debye screening, the electric field, or, equivalently, the time component of the vector potential (a three-dimensional scalar), becomes screened at a characteristic distance which can be understood as arising from massive virtual modes that have decoupled from the low-energy three-dimensional theory.

In Kaluza-Klein gravity, it has been shown[6] that the massive modes associated with the compact dimensions produce a quantum effective potential as a function of the metric associated with these dimensions. This potential can then force this metric field to diminish to a value such that the size of the extra dimensions is on the order of the Planck length. Thus quantum effects associated with the extra dimensions may provide the explanation for the smallness of these dimensions. This talk will be devoted to a description of this idea.

In the standard Abelian Kaluza-Klein model, one considers the usual Einstein action without cosmological term:

$$S = \frac{-1}{16\pi G_D} \int d^D x \, \sqrt{-g} \, R \tag{1}$$

where the scalar curvature R, defined from a dimensionless metric, has dimension $(\text{length})^{-2}$, and the D-dimensional gravitational constant has dimension $(\text{length})^{D-2}$. I will now take D to be five, and I will assume that the fifth dimension is finite in extent:

$$0 < x^5 < 2\pi R_5 \quad . \tag{2}$$

This means that the metric tensor can be expanded in a Fourier series

$$g_{\mu\nu}(x^i, x^5) = \sum_{n=-\infty}^{\infty} g^{(n)}(x^i) \, e^{inx^5/R_5} \quad . \tag{3}$$

The usual dimensional reduction consists in keeping only the n=0 mode. It is then conventional, and convenient, to parametrize the metric in the form:

$$g_{\mu\nu} = \phi^{-1/3} \begin{bmatrix} g_{ij} + A_i A_j \phi & A_i \phi \\ A_j \phi & \phi \end{bmatrix} \quad . \tag{4}$$

Then g_{ij} can be interpreted as the metric of four-dimensional space-time, A_i as the electromagnetic potential, and ϕ as a massless scalar field. The theory possesses the usual four-dimensional general covariance and Abelian gauge invariance of the coupled Einstein-Maxwell system. When $A_i=0$, the dynamics is equivalent to the Brans-Dicke scalar-tensor theory of gravity[7], with the Brans-Dicke parameter ω set to zero. The Weyl factor $\phi^{-1/3}$ is chosen so that the dimensionally-reduced action has the form

$$S_4 = -\frac{1}{16\pi G} \int d^4 x \, \sqrt{-\det g_{ij}} \left[R^{(4)} + \frac{1}{4} \phi F_{ij} F^{ij} + \frac{1}{6} g^{ij} \frac{\partial_i \phi \partial_j \phi}{\phi^2} \right] \tag{5}$$

where $R^{(4)}$ is the four-dimensional scalar curvature and $G=G_4=G_5/2\pi R_5$ is the four-dimensional gravitational constant. The Weyl factor is included only for convenience. Without it, a power of ϕ would have multiplied $R^{(4)}$, changing some of the bookkeeping but none of the physical conclusions.

By analogy with finite-temperature gauge theories, one might ask whether the scalar field acquires a mass due to one-loop corrections. In a Yang-Mills theory at finite temperature, the mass generation of $\vec{A}_0$ is already evident from the zero-mode contributions to the loops, where it appears with a linear ultraviolet divergence. When the non-zero modes are then included and summed over, the linear divergence is cut off by the temperature[5]. One can see immediately, however, that no such effect can be obtained simply by computing quantum corrections based on the zero-mode action S_4, because S_4 has the invariance

$$A_i \to \lambda A_i \qquad (6)$$

$$\phi \to \frac{1}{\lambda^2}\phi$$

which prevents the occurrence of a mass term for ϕ. More generally, the complete one-loop effective potential for ϕ, calculated using S_4, vanishes[8].

Thus the main computational goal is to obtain the effective potential for the scalar field, including the contribution of the massive modes. In performing this computation, it suffices to adopt a rather conservative view of Eq. (1) as a model theory of quantum gravity. That is, it can be viewed as an effective action which describes physics at distance scales larger than the Planck length. The scalar field itself is related to the (5,5) component of the metric; indeed the distance around the fifth dimension is

$$\int_0^{2\pi R_5} \phi^{1/3}\, dx^5 \quad .$$

Thus we expect that for $\phi^{1/3} R_5$ sufficiently large, our results should be reliable. This will be made more precise below.

In D dimensions, the gravitational field has D(D-3)/2 independent dynamical degrees of freedom. When D=5, the degrees of freedom described by $g^{(0)}_{\mu\nu}(x^i)$ in Eq. (3) are the massless graviton, photon and scalar. By contrast, the five degrees of freedom corresponding to $g^{(n)}_{\mu\nu}(x^i)$, $n \neq 0$, are those of a massive spin-two particle[9]. Thus the massive modes are purely spin two. It is convenient, in quantizing the theory, to choose a gauge condition which eliminates the unphysical massive modes associated with the vector and scalar fields:

$$g_{\mu 5,5}(x) = 0 \quad . \qquad (7)$$

I will call this the cylindrical gauge condition. It might seem the stronger condition $g_{\mu 5}(x)=0$ could be imposed, but this would be incompatible with the assumed periodicity of $g_{\mu\nu}$ in the x^5-direction. (This is directly analogous to finite-temperature gauge theories in which the static gauge $\partial_0 A_0=0$ is allowed, but temporal gauge, $A_0=0$, is in general incompatible with the periodicity of A_μ in time[10].)

The quantum theory is assumed to be given by the path integral

$$Z = \int Dg_{\mu\nu}\, \mu(g)\delta(g_{\mu 5,5})\Delta(g)e^{-S} \qquad (8)$$

where S is the action, Eq. (1), $\Delta(g)$ is the Faddeev-Popov ghost determinant corresponding to the gauge choice Eq. (7), and $\mu(g)$ is the measure of integration. Henceforth I shall assume $\mu=1$[11].

Before examining the effective potential of the scalar field, it is worth considering the general features of the loop expansion in quantum gravity. In D dimensions, the maximal degree of divergence d of an L-loop graph is independent of the number of external legs. It is given by

$$d = (D-2)L + 2 \quad . \tag{9}$$

I expect that the ultraviolet divergences of the Kaluza-Klein model will be the same as those of the uncompactified five-dimensional theory even though the integration over k^5 is replaced by the discrete mode sum[12]. Although nonrenormalizability is apparent from Eq. (9) with D=5, only those quantities which are divergent by naive power counting and for which counterterms are allowed by five-dimensional general covariance can avoid being finite. In five dimensions, a quintic ultraviolet divergence is the worst possibility at one loop, corresponding to a counterterm with no derivatives of the fields. In addition, there could be a cubic divergence associated with a two-derivative counterterm, and a linearly divergent four-derivative counterterm. Each of these is defined to be zero by the dimensional continuation or zeta-function regularization schemes, but for my purposes it is instructive to employ a dimensionful regulator for which they are non-vanishing[13].

Since my primary concern in this talk is the computation of the effective potential, I shall restrict my attention to possible zero-derivative terms. The only such counterterm which could be added to S (Eq. (1)) corresponds to an induced cosmological constant. It shall be possible to identify and separate off this cutoff-dependent, but R_5-independent, term, leaving a finite effective potential as part of the low-energy four-dimensional theory.

I now proceed to the computation of the effective potential at the one-loop level. I parametrize the fields as follows:

$$\begin{aligned} g_{ij} &= \eta_{ij} + h_{ij} \\ \phi &= \phi_c + \phi' \end{aligned} \tag{10}$$

and expand the action, keeping only terms that are quadratic in h_{ij}, A_i, and ϕ', but keeping all orders in ϕ_c, which can be regarded as a classical background field. For the purpose of computing the effective potential, it can be taken to be a constant. The resulting effective one-loop Lagrangian is

$$\begin{aligned} L_{\text{one-loop}} = -\frac{1}{16\pi G_5}\Bigg\{ & \frac{1}{6}\eta^{ij}\frac{\partial_i\phi'\partial_j\phi'}{\phi_e^2} - \frac{1}{4}\Big[-\eta^{ij}h^{k\ell}{}_{,i}h_{k\ell,j} + 2\eta_{ij}h^{ik}{}_{,\ell}h^{j\ell}{}_{,k} \\ & - 2\eta^{ij}h^{k\ell}{}_{,k}h_{ij,\ell} + \eta^{ij}\eta^{k\ell}\eta^{mn}h_{k\ell,i}h_{mn,j}\Big] \\ & - \frac{1}{4}\frac{1}{\phi_c}\Big[-h^{k\ell}{}_{,5}h_{k\ell,5} + \eta^{ij}\eta^{k\ell}h_{ij,5}h_{k\ell,5}\Big] \\ & + \frac{1}{4}\phi_c\eta^{ij}\eta^{k\ell}F_{i\ell}F_{jk}\Bigg\} \quad . \end{aligned} \tag{11}$$

Recall that ϕ' and A_i are x^5-independent in cylindrical gauge. Because $L_{one-loop}$ is quadratic in the quantum fields, the path integral for Z can be done exactly and the effective potential can be extracted using the formula[14]

$$Z = \exp\left\{- V_{eff}\sqrt{-g}\int d^5x\right\} . \tag{12}$$

Now observe that the ϕ_c dependence can be removed from the ϕ' and A_i terms simply by rescaling these fields. The same is true of the zero-mode piece of the h_{ij} terms. Therefore, these terms will make a contribution of the form $\delta^4(0)\ell n\phi_c$ to the logarithm of Z. These can, and should, be absorbed into the measure[11], and are not properly part of the effective potential. This is a restatement of the symmetry argument (Eq. (6)) that the zero-mode sector by itself does not generate an effective potential.

Incidentally, explicit computation reveals that the contribution of the Faddeev-Popov ghost to one-loop order is also of the same $\ell n\phi_c$ form, and hence does not contribute to the potential. I shall not discuss the ghost term further.

The only terms which do not share this scaling property are the massive modes. The kinetic terms are independent of ϕ_c, while the mass terms (i.e. those with x^5-derivatives) are proportional to $1/\phi_c$. I carry out the functional integral and drop terms in the exponent proportional to $\delta^5(0)\ell n\phi_c$ (for the same reasons as above). Using the definition of V_{eff} and the fact that $\sqrt{-g}=\phi_c^{-1/3}$, I find

$$\phi_c^{-1/3}\, V_{eff}(\phi_c) - V_{eff}(1) = \frac{1}{4\pi R_5}\int \frac{d^4k}{(2\pi)^4} \sum_{n=-\infty}^{\infty} 5\ell n \left[\frac{k^2 + \dfrac{n^2}{\phi_c R_5^2}}{k^2 + \dfrac{n^2}{R_5^2}}\right] \tag{13}$$

The subtraction, which normalizes the potential, has been made arbitrarily at $\phi_c=1$. In graphical terms, I have used $\phi_c=1$ to define the unperturbed, massive spin-two Lagrangian, and then summed all insertions proportional to $(1/\phi_c-1)$ on the one-loop graph, with the massive spin-two particles circulating around the loop.

To evaluate the potential, I use the formula

$$\sum_{n=-\infty}^{\infty} f_n = \int_{\infty}^{\infty} dz f(z) + \int_{-\infty+i\varepsilon}^{\infty+i\varepsilon} dz \frac{[f(z)+f(-z)]}{e^{-2\pi i z}-1} \tag{14}$$

where $f(z=n)=f_n$. I find

$$\phi_c^{-1/3} V_{eff}(\phi_c) - V_{eff}(\phi_c=1) = \frac{5}{2}\int \frac{d^5k}{(2\pi)^5}\, \ell n \left[\frac{k^2 + \frac{1}{\phi_c}k_5^2}{k^2+k_5^2}\right]$$
$$+ \frac{5}{(2\pi R_5)^5}\left[\frac{1}{\phi_c^2} - 1\right]\int \frac{d^4q}{(2\pi)^4}\, \ell n(1-e^{-q}) \tag{15}$$

(I have defined $k_5 = \frac{z}{R_5}$.)

As it stands, the first term is quntically divergent. However, even if a dimensionless regulator (which would define it to zero) is not employed, it still must properly be removed from the effective potential. To see this, let us recall the analogy between this computation and the Casimir effect[15]. In 1948, Casimir showed that the vacuum fluctuations of the electromagnetic field produce a small attractive force between two perfectly conducting parallel plates. One computes the zero-point energy of the field subject to the perfect-conductor boundary conditions, and subtracts from that the zero-point energy of the free electromagnetic field in the same volume of space.

The Kaluza-Klein problem is very similar. The "plates" in this case correspond to the boundaries $x^5=0$ and $x^5=2\pi R_5$, and the one-loop effective potential is just the zero-point energy density of the massive spin-two excitations confined to this "cavity". (Higher loop corrections would take into account the interactions of these spin-two particles with each other and with the massless modes.) The contribution from the first term in Eq. (15) is the energy density in uncompactified space. It is the precise analog of the term one subtracts out to obtain the measurable Casimir effect.

To see this more explicitly, I carry out the k_5 integral in this term. It is finite except for another $\ell n\phi_c$ term which we again drop. The remainder is proportional to $\sqrt{\phi_c}\int_0^\infty k^4 dk\Big|_{\phi_c=1}^{\phi_c}$. If a large sphere in momentum space is used to define this integral, then from the form of the metric (Eq. (6)), one sees that the radius should be taken to be $\phi_c^{-1/6}\Lambda$, in order that the cutoff Λ be independent of the choice of coordinates. The above expression then becomes $\frac{\Lambda^5}{5}\left|\frac{1}{\phi_c^{1/3}} - 1\right|$. Comparison with the left-hand side of Eq. (15) shows that this indeed represents a constant contribution to the effective potential - an induced cosmological constant.

The second term in Eq. (15) is finite. It is the physical effective potential, which I denote by $\tilde{V}_{eff}(\phi_c)$. Evaluating the integral, I have

$$\tilde{V}_{eff}(\phi_c) = -\frac{15}{4\pi^2}\zeta(5)\frac{1}{(2\pi R_5\phi_c^{1/3})^5} \tag{16}$$

where $\zeta(5) \cong 1.04$. This function is illustrated in Figure 1.

An effective potential which decreases to negative infinity as $\phi_c \to 0$ suggests a cataclysmic collapse of the fifth dimension. It is important, however, to estimate the range of validity of this one-loop result. Since G_5 is the loop expansion parameter, and since ϕ_c and R_5 can appear only in the combination $\phi_c^{1/3}R_5$, the dimensionless expansion parameter for the effective potential should be $G_5/\phi_c R_5^3$. Thus it is only for distances $2\pi\phi_c^{1/3}R_5$ around the fifth dimension, that are large compared to the Planck length $G_5^{1/3}$ [16] (the unshaded region in Fig. 1), that the result Eq. (16) is reliable. I conclude, then, that the vacuum fluctuations of

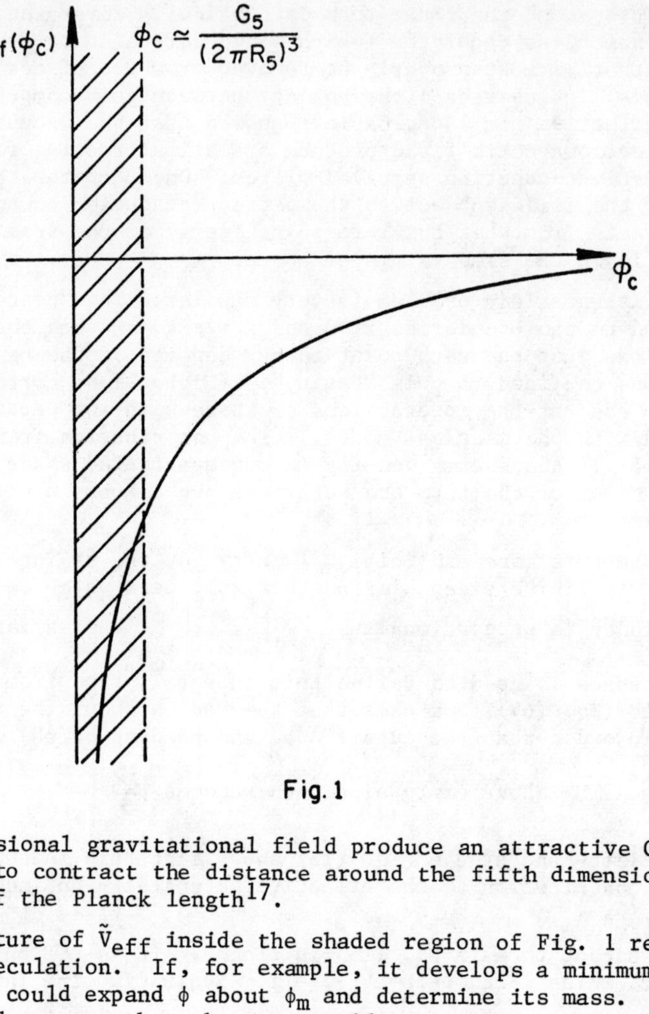

Fig. 1

the five-dimensional gravitational field produce an attractive Casimir force tending to contract the distance around the fifth dimension down to the order of the Planck length[17].

The structure of $\tilde{V}_{eff}$ inside the shaded region of Fig. 1 remains a matter of speculation. If, for example, it develops a minimum at $\phi=\phi_m$, then one could expand ϕ about ϕ_m and determine its mass. The scales involved suggest that the mass would be at least as large as the inverse Planck length, and therefore that ϕ would be "Debye screened" out of the low-energy four-dimensional theory (Eq. (5)). Furthermore, if the potential does pick a value ϕ_m of the scalar field in this way, then the distance around the fifth dimension will have been fixed relative to G_5. Thus the electric charge of the massive modes, given by an integral multiple of this ratio, will be determined.

The type of analysis I have described is not at all restricted to five dimensions. The most straightforward generalization is to a D-dimensional Einstein theory in which d(<D) of these dimensions are assumed to be compactified into circles. An effective potential as a function of the radii R_{D-d+1}, R_{D-d+2}, ... R_D can be constructed and then analyzed. This will determine which configurations are stable, with radii contracting to the Planck length, and which are unstable, with radii expanding out to infinitiy. A special case of this general analysis has

been reported by Rubin and Roth[18]. They compactified two of five dimensions, with one radius interpreted as inverse temperature 1/T, and considered the effective potential, at fixed temperature, as a function of R_5. They found that R_5 could be driven either to the Planck length or out to infinity depending on the relative size of R_5 and 1/T. The general analysis[19] of this kind of compactification is a first step in trying to answer the important question of whether quantum effects can determine the allowed, stable topologies in Kaluza-Klein theories.

Realistic theories based on non-Abelian gauge theories can be directly obtained by dimensional reduction only if the compact dimensions form a manifold with non-vanishing curvature[2]. The gauge coupling constant will appear as the ratio of the compact size to the Planck length and classical potentials will exist in the zero-mode sector, which can compete with the quantum effective potential. The inclusion of fermions is perhaps best considered within the framework of supersymmetric models. They offer control over the cosmological energy density and they might even be useful at distances below the Planck length.

REFERENCES AND FOOTNOTES

1. Th. Kaluza, Sitz. Preuss. Akad. Wiss. Berlin, Math. Phys. Kl, 966 (1921); O. Klein, Z. Phys. 37, 895 (1926).
2. For a sampling of recent contributions, see e.g.: A. Salam and J. Strathdee, Annals of Physics 141, 316 (1982); C.A. Orzalesi, Fortschritte der Physik 29, 413 (1981); E. Witten, Nucl. Phys. B186, 412 (1981); C. Wetterich, Phys. Lett. 113B, 377 (1982); S.D. Unwin, Phys. Lett. 103B, 18 (1981); P.G.O. Freund and M.A. Rubin, Phys. Lett. 97B, 233 (1980).
3. See, for example, E. Cremmer and B. Julia, Phys. Lett. 80B, 48 (1978); J. Sherk and J. Schwarz, Nucl. Phys. B153, 61 (1979).
4. N.S. Manton, Nucl. Phys. B158, 141 (1979) and B193, 502 (1981); G. Chapline and R. Slansky, Los Alamos report LA-UR-82-1076 (1982).
5. T. Appelquist and R. Pisarski, Phys. Rev. D23, 2305 (1981) and references contained therein.
6. T. Appelquist and A. Chodos, Phys. Rev. Lett. 50, 141 (1983).
7. C. Brans and R.H. Dicke, Phys. Rev. 124, 925 (1961).
8. This can be checked graphically as follows: Let $\phi=e^{\chi}$ and compute $V_{eff}(\chi)$. The integrals involved are quartically divergent but they can be regulated by introducing a large sphere in Euclidean momentum space. This respects the symmetry (Eq. (6)) without simply defining the integrals to vanish. It can then easily be seen that the graphs contributing to $V_{eff}(\chi)$ cancel. This can be done to any order in the loop expansion.
9. For a discussion of this point, see A. Salam and J. Strathdee, Ann. Phys. 141, 316 (1982), Appendix 5.
10. A recent analysis of finite temperature QCD in static gauge has been given by S. Nadkarni, Yale University preprint YTP82-21, August 1982.
11. G. 't Hooft in "Recent Developments in Gravitation" Cargèse 1978 (Plenum Press, New York, 1979), Especially section 3; B.S. DeWitt In "General Relativity, an Einstein Centenary Survey", S.W. Hawking and W. Israel, eds. (Cambridge University Press, Cambridge UK, 1979), especially sec. 14.5.2 and pp. 734-735.

12. This intuitive expectation is certainly born out by my explicit computations. In finite temperature gauge theories, it has been proven through at least two orders in the loop expansion. For a recent discussion of infinities in Kaluza-Klein theories, see M. Duff and D. Toms, CERN Report TH.3259-CERN, March 1982.
13. A simple cutoff, which allows one to define and separate off the unobservable, infinite addition to the effective potential, is a large sphere in Euclidean momentum space (see Footnote 8).
14. R. Jackiw, Phys. Rev. D9, 1686 (1974). The explicit factor of $\sqrt{-g}$ is included so that the one-loop V_{eff} will be generally covariant.
15. H.B.G. Casimir, Koninkl. Ned. Akad. Wetenschap. Proc. Ser. B, 793 (1948).
16. This is the "Planck length" seen by a five-dimensional observer. The usual Planck length, defined e.g. in terrestrial measurements, is $\tilde{G}_4=G_5/(2\pi R_5\phi_c^{1/3})$. However, since I have chosen coordinates scaled by $\phi_c^{1/6}$ (see Eq. (4)), the parameter that multiplies the lagrangian (Eq. 5) is $\phi_c^{1/3}\tilde{G}_4=G_4$.
17. E. Witten [Nucl. Phys. B195, 481 (1982)] has argued that the five-dimensional Kaluza-Klein vacuum is semiclasically unstable. This result, however, depends on the specific model. By contrast, I expect that the present analysis can be straightforwardly generalized to any Kaluza-Klein model of interest.
18. M. Rubin and B. Roth, University of Texas preprint, January 1983.
19. T. Appelquist, A. Chodos and E. Myers, Yale University, Manuscript in preparation.

SUPERGRAVITY AND UNIFICATION*

R. Arnowitt†
Lyman Laboratory of Physics
Harvard University
Cambridge, MA 02138

and

A.H. Chamseddine and Pran Nath
Department of Physics
Northeastern University
Boston, MA 02115

ABSTRACT

A survey of properties of grand unification schemes based on N=1 Supergravity is given. The topics discussed include the following: the construction of Supergravity GUT models, spontaneous symmetry breaking of Supergravity and SU(2)xU(1) symmetry, general conditions at the tree level for the existance of a low mass sector "protected" from the GUT and Planck masses, construction of the low mass effective potential, and conditions for maintaining the gauge hierarchy up to the one loop level. Properties of a general class of models possessing a light photino and a light gluino are discussed, and some features of the supersymmetric decays of the W and Z mesons are described.

*Research is supported in part by the National Science Foundation under Grant No. PHY 77-22864 and Grant No. PHY 80-0833.

† On Sabbatical leave from Department of Physics, Northeastern University, Boston, MA 02115

I. INTRODUCTION

Supergravity grand unification is an attempt to combine N=1 supergravity with conventional grand unified models. Supergravity itself is concerned with supersymmetric gravitational interactions and is governed by the Planck mass $M_{PL} = \kappa^{-1}$ where

$$\kappa = (8\pi G)^{1/2} = 0.41 \times 10^{-18}\ \text{GeV}^{-1} \tag{1.1}$$

Standard GUT models, on the other hand deal with the rest of physics, strong, electroweak and superheavy physics, and are governed by the electroweak mass scale m_{e-w} and the GUT scale M, where

$$m_{e-w} \sim 100\ \text{GeV} \tag{1.2a}$$

$$M \sim 10^{16}\ \text{GeV} \tag{1.2b}$$

The gauge group used to combine these two theories is the product group

$$(\text{N=1 Supergravity}) \times G \tag{1.3}$$

where G is a conventional grand unified group (e.g. G = SU(5), O(10), etc.).

At first sight it might appear that such a union of supergravity and GUT models is unmotivated. Thus the mass scales of the two theories appear disparate. Each theory pollutes the other with its illnesses! Thus the non-renormalizable nature of supergravity matter couplings infect the previously renormalizable GUT models, while the arbitrariness of the GUT Higgs couplings effect the previously rigid supergravity couplings. Even the gauge group (1.3) is a product group. In spite of this, a union of these ideas does appear to arise from the symmetry breaking occurring in the theory. In particular, there exist models where the spontaneous breaking of supergravity and the spontaneous breaking of SU(2)xU(1) arise from a common source scaled by an "intermediate mass" m. In particular, the gravitino and W mesons, whose masses m_g and M_W govern the breaking of supergravity and SU(2)xU(1) respectively, grow masses of size

$$m_s \equiv \kappa m^2 \tag{1.4}$$

i.e., both $m_g \sim m_s$ and $M_W \sim m_s$. From Eq. (1.2a) one has $m_s \sim 100$ GeV, from which one concludes that the intermediate mass m is of size

$$m \sim 10^{10} \text{ GeV} \tag{1.5}$$

Also since $m_g \sim M_W$, one has the remarkable result that effectively supergravity is a good symmetry down to the electroweak mass scale. While models proposed up to now are not fully satisfactory, the possibility that Supergravity may shed light on SU(2)xU(1) breaking makes Supergravity GUT models an interesting area of investigation.

Any theory that includes gravity into the unification, possesses, however, a new gauge hierarchy problem. This is due to the closeness of the GUT and Planck masses, i.e., $\kappa M \sim 10^{-2}$. Thus, if to lowest order the electroweak scale is of order 100 GeV, gravitational corrections to m_{e-w} of size

$$(\kappa M)M;\ (\kappa M)^2M;\ .\ .\ .\ (\kappa M)^6M \tag{1.6}$$

must all be suppressed. Thus the low mass electroweak sector of the theory must now be "protected" from both the GUT scale M and the Planck scale κ^{-1}. How this protection is achieved (a problem even at the tree level) is an important question in models of this type.

The topics we will consider in this review are the following: (1) A very brief description of the construction of Supergravity GUT interactions; (2) A discussion of the general class of models possessing the "protection" of the low energy sector; (3) A description of some of the properties of the low mass sector (mass matrices, three point interactions) for a class of models; and (4) Some phenomenological predictions concerning supersymmetric decay channels of the W and Z for these models.

II. CONSTRUCTION OF SUPERGRAVITY GUT LAGRANGIANS

As in global SUSY models, the quarks, leptons and Higgs scalars of Supergravity GUT models are members of left-handed chiral scalar multiplets Σ_A, where

$$\Sigma_A = (Z_A, X^A_L, h_A);\ A = 1\ .\ .\ .\ N \tag{2.1}$$

Here the index A enumerates the multiplets (and includes the G group index), Z_A are complex scalar fields, X^A_L l.h. spinors and h_A complex auxillary fields. Thus for the quark/lepton multiplets, X_L^A represent the quark/lepton Weyl spinors and Z_A are the supersymmetric scalar partners (squarks and sleptons). For the Higgs multiplets, Z_A are the Higgs scalars and X_L^A their spinor partners (Higgsinos).

The gauge mesons of G are embedded in the vector multiplet, which has the general form

$$V = (C, \xi, H, K; V_\mu, \lambda, D) \tag{2.2}$$

where V_μ is a vector field, C, H, K and D are scalar fields, and ξ and λ are Majorana spinors. The gauge vector multiplet in the Wess-Zumino gauge reduces to the simple form

$$V^\alpha = (0, 0, 0, 0; V_\mu^{\ \alpha}, \lambda^\alpha, D^\alpha) \tag{2.3}$$

where α is the adjoint representation G index, $V_\mu^{\ \alpha}$ is the gauge vector meson fields, λ^α are the Majorana spinor superpartners (gauginos) and D^α are scalar auxilliary fields.

For minimal auxilliary variables the N=1 Supergravity multiplet is given by

$$S.G. = (e_\mu^{\ a}, \psi_\mu; A_\mu, u = S-iP) \tag{2.4}$$

where $e_\mu^{\ a}$ is the vierbein field, ψ_μ the gravitino, and A and u are auxilliary fields. The Supergravity Lagrangian is then [1]

$$L_{S.G.} = -\frac{3}{2\kappa^2} R(e,\omega) - \frac{1}{2} \varepsilon^{\mu\nu\rho\tau}\psi_\mu\gamma_5 \gamma_\nu D_\rho\psi_\sigma$$

$$- \frac{1}{3} e|u|^2 + \frac{1}{3} e A_\mu A^\mu; \quad e \equiv \det e_\mu^{\ a} \tag{2.5}$$

where R is the curvature scalar.

The rules for coupling scalar and vector multiplets to N=1 Supergravity have been known for some time [2]. Thus the Supergravity invariant Lagrangian coupling a scalar multiplet reads

$$e^{-1}L_{\Sigma} = \frac{1}{2}\,[h + \kappa uZ + \kappa\bar{\psi}_{\mu}\gamma^{\mu}\chi_{L} + \kappa^{2}\bar{\psi}_{\mu}\sigma^{\mu\nu}\psi_{\nu R} + \text{h.c.}] \qquad (2.6a)$$

while for the general vector multiplet the invariant Lagrangian is

$$e^{-1}L_{V} = D - \frac{i\kappa}{2}\,\psi_{\mu}\gamma_{5}\gamma^{\mu}\lambda - \frac{\kappa}{3}\,(u^{*}v + uv^{*}) + \frac{2}{3}\,\kappa A_{\mu}V^{\mu}$$

$$+ i\,\frac{\kappa^{2}}{4}\,e^{-1}\,\varepsilon^{\mu\nu\rho\sigma}\bar{\psi}_{\mu}\gamma_{\nu}\psi_{\rho}V_{\sigma} + \frac{i\kappa}{3}\,e^{-1}\bar{\xi}\gamma_{5}\gamma_{\mu}R^{\mu}$$

$$+ i\,\frac{\kappa^{2}}{8}\,e^{-1}\,\varepsilon^{\mu\nu\rho\sigma}(\psi_{\mu}\gamma_{\nu}\psi_{\rho})(\bar{\xi}\psi_{\sigma}) - \frac{2}{3}\,\kappa^{2}e^{-1}\,C\,L_{S.G.} \qquad (2.6b)$$

The most general Lagrangian coupling a single scalar multiplet to Supergravity was worked out previously by Cremmer *et al.* [3]. For a Supergravity GUT model one must deal with many scalar multiplets as well as the gauge vector multiplet, and guarantee that all results are both Supergravity *and* G invariant. This generalization has been carried out independently by several groups [4,5,6]. To summarize briefly, to obtain the most general couplings with gauge invariance (1.3), one may proceed as follows: (1) First couple the gauge vector multiplet V^{α} to the chiral multiplets Σ_a in the most general G-invariant way. As in global SUSY theories, this leads to a G-invariant vector multiplet. Then couple this vector multiplet to Supergravity using Eq. (2.6b). (2) Couple the Σ_a multiplets together in the most general G-invariant way. As in global SUSY theories, this leads to a G-invariant chiral scalar multiplet. Then couple this chiral multiplet to Supergravity using Eq. (2.6a). The total Lagrangian is thus

$$L = L_{S.G.} + L_V + L_\Sigma + L_{Y.M.} \tag{2.7}$$

where $L_{Y.M.}$ is the supersymmetric Yang-Mills Lagrangian for the gauge multiplet (2.3) [7]. (3) Then eliminate all auxilliary variables and simplify. One obtains in this fashion the most general Lagrangian (with at most two derivatives) possessing the gauge invariance (1.3) and explicitly expressed in terms of the independent dynamical fields.

The final form for L depends on two arbitrary functions of the scalar fields Z_a: a function $f_{\alpha\beta}(Z_a)$ which transforms as the symmetric product of adjoint representations of G and a G-invariant function

$$\mathcal{G}(Z_A, Z_A^+) = 3\ln[-\frac{\kappa^2}{2}\phi(Z_1Z^+)] - \ln\frac{\kappa^6}{4}|g(Z)|^2 \tag{2.8}$$

where

$$g(Z_A) = \text{superpotential} \tag{2.9}$$

$f_{\alpha\beta}$ enters in the Yang-Mills kinetic energy, $-\frac{1}{4}f_{\alpha\beta}F_{\mu\nu}^{\alpha}F_{\mu\nu}^{\beta}$, and ϕ enters in the scalar field kinetic energy:

$$[\frac{3}{\kappa^2}\partial^2(\ln\phi)/\partial Z_A\partial Z_B^+]\partial_\mu Z_A^+\partial^\mu Z_B \tag{2.10}$$

The superpotential plays the same role in Supergravity GUT models as it did in the global SUSY case. Clearly the choices

$$f_{\alpha\beta} = \delta_{\alpha\beta} \tag{2.11a}$$

$$\phi(Z\,Z^+) = -\frac{3}{\kappa 2}\exp[-\frac{\kappa^2}{6}\sum Z_A^+Z_A] \tag{2.11b}$$

reduce the Yang-Mills and scalar kinetic energies to canonical form. In global SUSY models, Eqs. (2.11) are always imposed to maintain renormalizability. Since Supergravity-matter interaction are already non-renormalizable, the imposition of this constraint is no longer obviously necessary. Weinberg [8] has argued that gravitational loop corrections should possess a U(N) symmetry in the matter fields to a good approximation, and that this symmetry would imply the validity of Eq. (2.11a) while Eq. (2.11b) would be replaced by the more general form

$$\phi(Z_1 Z^+) = F(\kappa^2 \sum_A Z_A{}^+ Z_A) \qquad (2.12)$$

In the discussion below, Eqs. (2.) and (2.) yield qualitatively similar predictions in the low energy domain and so, for simplicity, we will assume Eqs. (2.11) hold. (One may easily extend the following results to the more general case of Eq. (2.) using the discussion of Ref. [9].)

III. TREE EFFECTIVE POTENTIAL

The Supergravity GUT Lagrangian described in the preceeding section yields the following effective potential in the tree approximation [4,5,6]

$$V = \frac{1}{2} \exp[\frac{\kappa^2}{2} Z_B{}^+ Z_B][G_A{}^+ G_A - \frac{3}{2} \kappa^2 |g(Z)|^2]$$

$$+ \frac{1}{32} [e_\alpha Z_A{}^+ (T^\alpha Z)_A]^2 \qquad (3.1)$$

where

$$G_A = \frac{\partial g(Z)}{\partial Z_A} + \frac{\kappa^2}{2} Z_A^+ g(Z) \tag{3.2}$$

Here T^α are the G-group generators and e_α the associated coupling constants. The tree level VEVs are determined from the extremen conditions

$$\frac{\partial V}{\partial Z_A} = 0 = \frac{\partial V}{\partial Z_A^+} \tag{3.3}$$

which on the real manifold ($\langle Z_A \rangle = \langle Z_B^+ \rangle$) may be reduced to the matrix equations

$$T_{AB} G_B = 0 \tag{3.4}$$

where

$$T_{AB} = g_{,AB} + \frac{\kappa^2}{2} (Z_A g_{,B} + Z_B g_{,A}) + \frac{\kappa^4}{4} Z_A Z_B g - \kappa^2 \delta_{AB} g \tag{3.5}$$

and $g_{,A} \equiv \partial g / \partial Z_A$ etc.

The condition that Supergravity invariance remain unbroken is that all the G_A vanish at the minimum, while if one or more of the G_A are non-zero, Supergravity invariance is broken. We consider now two simple examples to illustrate how Supergravity and G invariance may be spontaneously broken at the tree level.

(1) Breaking of Supersymmetry (Super Higgs Effect)

It is generally difficult to break global supersymmetry spontaneously and the usual mechanisms used require at least three superfields. In contrast, local Supergravity symmetry can be broken in a straight forward way using a supersymmetric generalization of the Higgs effect [10,3]. If we introduce a single chiral, G-invariant superfield $\sum_Z \equiv (Z, X_L, h_Z)$, the simplest possible superpotential one might consider is a linear one

$$g = g_2 \equiv m^2(Z + B) \tag{3.6}$$

where m and B are constants with dimensions of mass. One may then verify that Eqs. (3.4) has solutions which break Supergravity invariance provided $B^2 < 8$. Adjusting B so that the cosmological constant is zero ($V_{min} = 0$) one finds

$$<\kappa Z> = \sqrt{6} - \sqrt{2}\ , \quad \kappa B = (2\sqrt{2} - \sqrt{6}) \tag{3.7}$$

while at the minimum

$$(g_2)_{min} = \sqrt{2}\ \frac{m^2}{\kappa}\ ;\ (G_Z)_{min} = \sqrt{3}\ m^2 \tag{3.8}$$

The fact that G_Z is non-zero is the signal that Supergravity has been broken spontaneously. A standard Higgs phenomena occurs [3]: X_L is the goldstino which is absorbed by the gravitino which then becomes massive. It is interesting to note that while the VEV of Z is of order $\kappa^{-1} = M_{PL}$, the mass associated with the Z field is very small,

$$m_Z = \kappa m^2 \equiv m_s \tag{3.9}$$

as is the gravitino mass (i.e., $m_g \sim m_s$).

(2) Breaking of Supersymmetry and G-Invariance

In all models examined so far, the breaking of the gauge group G to SU(3)xSU(2)xU(1) is accomplished at the tree level, much as in global SUSY models. Examples exist in which the further breaking of SU(2)xU(1) to $U_{E.M.}(1)$ occurs at the tree level or alternately at higher loop order. In either case it is the Supergravity breaking parameter $m_s = \kappa m^2$ that scales the SU(2)xU(1) breaking. We discuss here only a very simple model [4] of tree breaking of SU(2)xU(1) to illustrate some of the ideas that occur [11].

We chose G = SU(5) and the superpotential to be $g = g_1 + g_2$ where g_2 has the simple Polony form Eq. (3.6) and g_1 carries the physics of the GUT sector:

$$g_1 = \lambda_1[\tfrac{1}{3}\,Tr\Sigma^3 + \tfrac{1}{2}\,MTr\Sigma^2] + \lambda_2 H_x'(\Sigma^x{}_y + 3M\delta^x{}_y)H^y$$

$$+ \lambda_3 U H_x' H^x - 5\lambda_1 M^3 \tag{3.10}$$

Here M is the GUT mass, $\Sigma^x{}_y$, H^x, H_x' and U are 24, 5, $\bar{5}$ and singlet representations of SU(5) [12]. The full set of scalar fields $\{Z_A\}$ consists of $\{Z_a\}$, the GUT fields in g_1, and the super Higgs field Z of Eq. (3.6). In order to solve the extremen conditions Eq. (3.4), we expand Z_A in a power series in κ:

$$Z_a = Z_a{}^{(0)} + \kappa Z_a{}^{(1)} + \ldots\,;\quad Z = \frac{1}{\kappa}\,Z^{(-1)} + Z^{(0)} + \kappa Z^{(1)} + \ldots \tag{3.11}$$

The solutions of Eq. (3.4) shows [4] that the matter fields $\{Z_a\}$ divide into a light set $\{L_\alpha\}$ and a heavy set $\{H_i\}$ where

$$\{L_\alpha\} = \{U;\ H_x',\ H^x,\ x = 4,5\} \tag{3.12}$$

$$\{H_i\} = \{\Sigma^x{}_y;\ H_x',\ H^x,\ x = 1,2,3\} \tag{3.13}$$

The masses and VEVs of the fields are

$$m_{H_i} \sim M;\ \ m_{L_\alpha} \sim \kappa m^2 \equiv m_s;\ \ m_Z \sim m_s \tag{3.14a}$$

$$\langle H_i\rangle \sim M;\ \langle L_\alpha\rangle \sim m_s;\ \ \langle Z\rangle \sim \kappa^{-1} \tag{3.14b}$$

The existance of the heavy VEVs produces the breaking of SU(5) to SU(3)xSU(2)xU(1) and the existance of the light VEVs signals the further spontaneous breaking of SU(2)xU(1). (The non-zero $\langle Z\rangle$ again implies the spontaneous breaking of supersymmetry.) The example is "realistic" in that the low energy sector correctly accounts for all SU(3)xSU(2)xU(1) physics when quark and lepton terms are added to g_1 [4].

Of course the above model is not really satisfactory. One had to dial the GUT mass parameters in the λ_1 and λ_2 couplings of Eq. (3.10) to be equal (as in the theory's global SUSY predecessor [12]). Also, the model has to be modified to maintain the separation between heavy and light fields at the loop level (see Sec. 6 below). However, it is remarkable that one can set up a model where at least at the tree level the light fields are "protected" from both the GUT mass M and the Planck mass $\kappa^{-1} \equiv M_{PL}$ to <u>all</u> orders in κ.

IV. GENERAL CONDITIONS FOR PROTECTION OF LOW MASS SECTOR

The existance of a tree level mass hierarchy, with a low mass SU(3)xSU(2)xU(1) sector protected from the large GUT and Planck masses, is actually characteristic of a wide class of models. In this section we discuss the general conditions needed to guarantee this mass separation, and briefly discuss the origin of the effect.

We consider a general superpotential of the form

$$g(Z_a,Z) = g_1(Z_a) + g_2(Z) \tag{4.1}$$

where $\{Z_a\}$ are the GUT fields and Z is the super Higgs field. We write g_2 in the general form

$$g_2 = \frac{m^2}{\kappa} f_2(\kappa Z) \tag{4.2}$$

and assume only that g_2 produces a spontaneous breaking of Supergravity with $\langle\kappa Z\rangle \sim O(1)$. [Thus for the simple Polony form of Eq. (6), one has $f_2(x) = x + \kappa B$ and the result of Eq. (3.7).] We assume a priori that g_1 depends only on the GUT mass M, which scales the breaking of G to SU(3)xSU(2)xU(1). We wish to determine the conditions on $g_1(Z_a)$ such that the VEVs of $\{Z_a\}$ deduced from the extremen equations (3.4) divide into a heavy and light fields, i.e., $\{Z_a\} = \{H_i, L_\alpha\}$ with $m_{H_i} = O(M)$ and $m_{L_\alpha} = O(m_s)$ where $m_s \equiv \kappa m^2$.

We first state the basic result [13]: Let $\{H_i\}$ be the fields with non-singular mass matrix M_{ij} whose eigenvalues are all O(M) at $\kappa = 0$, and $\{L_\alpha\}$ the remaining fields. If at the minimum of V,

$$g_{,\alpha i} \equiv \partial^2 g_1/\partial L_\alpha \partial H_i \sim O(m_s) \tag{4.3a}$$

$$g_{,\alpha\beta} \equiv \partial^2 g_1/\partial L_\alpha \partial L_\beta \sim O(m_s), \tag{4.3b}$$

Then the solutions of Eqs. (3.4) possess expansions of the form

$$L_\alpha = O(m_s) + O(m_s^2/M) + O(\kappa^2 M m_s) + \ldots \tag{4.4a}$$

$$H_i = O(M) + O(m_s) + O(m_s^2/M) + \ldots \tag{4.4b}$$

$$\kappa Z = O(1) + O(m_s/M) + O(\kappa^2 M m_s) + \ldots \tag{4.4c}$$

while the corresponding masses are

$$m_\alpha^2 = O(m_s^2) + O(m_s^3/M) + \ldots \tag{4.5a}$$

$$M_L^2 = O(M^2) + O(m_s M) + \ldots \tag{4.5b}$$

$$m_Z^2 = O(m_s^2) + O(m_s^3/M) + \ldots \tag{4.5c}$$

where the + . . ., terms have increasing powers of M or $M_{PL} \equiv \kappa^{-1}$ in the denominator. Further, if one rescales the functions G_A of Eq. (3.2) by

$$G_i = m_s^2 \bar{G}_i; \; G_\alpha = m_s^2 \bar{G}_\alpha; \; G_Z = m^2 \bar{G}_Z \tag{4.6}$$

then all the $\bar{G}_A$ are O(1) <u>at the minimum</u>.

Eqs. (4.3a) are the basic conditions on $g_1(Z_a)$ to maintain a tree level mass hierarchy to all orders in κ. Clearly a wide class of models is allowed, though specific couplings are excluded. For example, Eq. (4.3a) implies that a term in g_1 of the form $\lambda_{ij\alpha} H_i H_j L_\alpha$ must have the coupling constant $\lambda_{ij\alpha} \sim m_s/M \sim 10^{-14}$ unless the first term in Eq. (4.4b) should happen to be zero. (Thus, color conservation implies that for the Higgs 5 representation $\langle H^x \rangle$ vanishes for x = 1,2,3.) Central to deducing the result is the additive form of the superpotential Eq. (4.1) [14]. This implies a mysterious separation between the Super-Higgs world and the rest of physics.

Eqs. (4.6) will be seen below to be important in allowing the gauge hierarchy to be maintained at the one loop level. Particularly remarkable is the result in the heavy sector for G_i. Recall that the vanishing of all the G_A corresponds to the situation of no Supergravity breaking. Away from the minimum one characteristically has that $G_i \sim O(M^2)$. Hence Eq. (4.6) shows that <u>at</u> the minimum G_i is reduced by a factor $(m_s/M)^2 \sim 10^{-28}$! (Similarly G_α is reduced by $m_s/M \sim 10^{-14}$.) Thus the mass hierarchy implies that the minimum is very deep and that supergravity is broken very gently.

We briefly sketch now the derivation of the above results. (A detailed discussion is given in [13].) It is convenient to first scale the fields by their expected sizes at the minimum, i.e., we write $g = (m^2/\kappa)\bar{g}$ and

$$H_i = M\bar{H}_i, \; L_\alpha = m_s\bar{L}_\alpha, \; Z = \bar{Z}/\kappa \tag{4.7}$$

The extremen equations (3.4) for the heavy fields, $A = H_i$ then reads $T_{iB}G_B = 0$ or

$$T_{ij}G_j + T_{i\alpha}G_\alpha + T_{iZ}G_Z = 0 \tag{4.8}$$

Inserting in Eqs. (3.2) and (3.5) gives

$$[\frac{1}{M}\, g_{,ij}\bar{G}_j] + [\frac{1}{2}\,\bar{H}_i\bar{G}_Z{}^2 - \frac{1}{4}\,H_i\bar{Z}\bar{g}\bar{G}_Z] + \frac{1}{M}\, g_{,i\alpha}\bar{G}_\alpha$$

$$+ \frac{m_s}{M}\,[\frac{1}{2}\,\bar{Z}\bar{G}_Z - \bar{g}]\bar{G}_i - \frac{1}{2}\,(\kappa M\kappa m_s)\bar{H}_i\bar{H}_j\bar{g}\bar{G}_j + \ldots = 0 \tag{4.9}$$

where the omitted terms are all of $O[(\kappa m_s)^2]$ and hence will be quite small. Similarly for the light sector ($A = L_\alpha$) and the super-Higgs sector ($A = Z$) one finds

$$\left[\boxed{\frac{1}{m_s}\bar{g}_{,\alpha\beta}} - \delta_{\alpha\beta}\left(\frac{1}{2}\bar{Z}\bar{G}_Z - \bar{g}\right)\right]\bar{G}_\beta + \left[\boxed{\frac{1}{m_s}\bar{g}_{,\alpha i}\bar{G}_i} + \frac{1}{2}\bar{L}_\alpha \bar{G}_Z^2 - \frac{1}{4}\bar{L}_\alpha \bar{Z}\bar{g}\bar{G}_Z\right] + \frac{1}{4}(\kappa M \kappa m_s)\left[2\bar{H}_i\bar{G}_i\bar{G}_\alpha - \bar{L}_\alpha\bar{H}_i\bar{G}_i\bar{g}\right] + \ldots = 0 \tag{4.10}$$

and

$$[\bar{g}_{,\bar{Z}\bar{Z}}] + [\bar{Z}\bar{G}_Z - \frac{1}{4}\bar{Z}^2\bar{g}] + \frac{1}{4}(\kappa M \kappa m_s)[2\bar{H}_i\bar{G}_i\bar{G}_Z + \bar{Z}\bar{H}_i\bar{G}_i\bar{g}] + \ldots = 0 \tag{4.11}$$

Eqs. (4.9 - 4.11) allow one to solve for the VEVs of all the fields. By hypothesis, $g_{,ij}$ is non-singular and $O(M)$. Thus the only terms that might violate the a priori size estimates of the VEVs are the boxed terms in Eq. (4.10) as they contain $1/m_s$ factors. Conditions (4.3) then are sufficient to allow expansions of form Eqs. (4.4) to be made in the extremen equations. The expansions for the masses of the fields, Eq. (4.5) arise in a similar fashion.

V. LOW ENERGY EFFECTIVE POTENTIAL

In any model which has a protected low energy sector, it should be possible to eliminate the heavy and super-Higgs fields to obtain a reduced or effective low energy theory. Thus Eqs. (4.9) and (4.11) can be used to express H_i and Z in terms of the light fields L_α in the full effective potential Eq. (3.1), to obtain a low energy effective potential which is a function only of L_α:

$$U(L_\alpha, L_\alpha^+) = V[L_\alpha; H_i(L_\alpha); Z(L_\alpha)] \tag{5.1}$$

The general construction of U up to terms of $O(m_s^4)$ was first carried out by Hall, Lykken and Weinberg [9]. We describe briefly here an alternate derivation discussed in Ref. [13] based on the protection theorem equations of Sec. IV [15].

Rather than calculating the low energy effective potential directly from Eq. (5.1), one may first use Eqs. (4.9) and (4.11) to solve for H_i and Z and insert these solutions into the low energy sector equations Eq. (4.10). The resultant form of Eq. (4.10) depends only on L_α and must be what one would obtain from extremizing the U of Eq. (5.1) and hence must be in the form of a gradient, i.e., $\partial U/\partial L_\alpha$. One can then integrate this expression to obtain U directly. To obtain the leading $O(m_s^4)$ part of U is simple, as one may then neglect all $\kappa^2 M m_s$, $(\kappa m_s)^2$ or m_s/M corrections, i.e., keep only the first two brackets of Eqs. (4.9) - (4.11) [16]. Eliminating H_i and Z then leads to the following result:

$$U(L_\alpha, L_\alpha^+) = \frac{1}{2} E_o[|\tilde{g}_{1,\alpha}|^2 + m_1^2 L_\alpha^+ L_\alpha + (\omega + \omega^+) + m_s^2(\tilde{g}_{1,i}\overline{G}_i^{(0)} + h.c.)] + \frac{1}{32}[e_\alpha L^+ T^\alpha L]^2 \tag{5.2}$$

where

$$\omega = m_2\tilde{g}_1 + m_3 L_\alpha g_{1,\alpha};\ E_o \equiv \exp \frac{1}{2}|\overline{Z}^{(0)}|^2 \tag{5.3}$$

$$\tilde{g}_1(H_i, L_\alpha) = g_1(H_i, L_\alpha) - g_1(H_i, 0) - b \tag{5.4}$$

and

$$m_1{}^2 = \frac{1}{2} m_s{}^2[|\bar{G}_Z{}^{(0)}|^2 - |\bar{g}_2{}^{(0)}|^2] = |m_3| \tag{5.5a}$$

$$m_2{}^2 = \frac{1}{2} m_s[\bar{Z}^{(0)}\bar{G}_Z{}^{(0)} - 3\bar{g}_2{}^{(0)}] \tag{5.5b}$$

$$m_3 = \frac{1}{2} m_s \bar{g}_2{}^{(0)} \tag{5.5c}$$

The heavy fields H_i and $\bar{G}_i{}^{(0)}$ are given by

$$H_i = H_i{}^{(0)} - \frac{1}{2} m_s \bar{g}_2{}^{(0)} (M^{-1})_{ij} H_j{}^{(0)} \tag{5.6a}$$

$$\bar{G}_i{}^{(0)} = \frac{1}{4}(M^{-1})_{ij} Z_j{}^{(0)} \bar{G}_Z{}^{(0)} [\bar{Z}^{(0)} \bar{g}_2{}^{(0)} - 2\bar{G}_Z{}^{(0)}] \tag{5.6b}$$

Here $M_{ij} \equiv g_{,ij}{}^{(0)}$ is the heavy sector's mass matrix and the constant $b \sim O(m_s{}^3)$ in Eq. (5.4) is to be chosen so that the cosmological constant vanishes.

The low energy effective potential is particularly useful tool for comparing Supergravity GUT models with experiment. We will make use of it in the discussion of the phenomenological predictions of a class of models in Sec. VII below [17].

VI. LOOP CORRECTIONS TO THE GAUGE HIERARCHY

The discussion up to now has concerned itself with the tree level gauge hierarchy. New problems arise at the loop level. The one loop effective potential $V^{(1)}$ for a supergravity GUT model [18] can be expressed as a supertrace (STr):

$$V^{(1)} = \frac{1}{64\pi^2} \mathrm{STr}[M^4 \ln(M^2/\mu^2)]$$

$$= \frac{1}{64\pi^2} \sum_{J=0}^{3/2} (-1)^{2J}(2J + 1)\mathrm{Tr}M_J^4 \ln(M_J^2/\mu^2) \qquad (6.1)$$

where M_J^2 is the mass matrix for particles of spin J and μ is the subtraction point. The spin 0 mass matrices of the scalar fields Z_A can be obtained directly from the tree effective potential V^{tree} of Eq. (3.1). Writing Z = A + iB, the A and B mass matrices are gotten from

$$(M_{(A)}^2)_{AB} = \left[\left(\frac{\partial}{\partial Z_A} + \frac{\partial}{\partial Z_A^+}\right)\left(\frac{\partial}{\partial Z_B} + \frac{\partial}{\partial Z_B^+}\right)V^{tree}(Z,Z^+)\right]_{\langle Z\rangle = \langle Z^+\rangle} \qquad (6.2)$$

$$(M_{(B)}^2)_{AB} = -\left[\left(\frac{\partial}{\partial Z_A} - \frac{\partial}{\partial Z_A^+}\right)\left(\frac{\partial}{\partial Z_B} - \frac{\partial}{\partial Z_B^+}\right)V^{tree}(Z,Z^+)\right]_{\langle Z\rangle = \langle Z^+\rangle} \qquad (6.3)$$

where one evaluates the r.h.s. of Eqs. (6.2), (6.3) at the tree values of the VEV. The spin 1/2 mass matrices for the chiral super partners, X_A, of the Z_A and the gaugino partners, λ^α, of the gauge mesons enter the Lagrangian as

$$-L_{1/2}^{mass} = \frac{1}{2} \bar{X}_A m_{AB} X_B + X_A \mu_{A\alpha} \lambda^\alpha \qquad (6.4)$$

where

$$m_{AB} = \tilde{g}_{,AB} - \frac{\kappa^2}{2} \delta_{AB}\tilde{g} - \frac{2}{3} \tilde{g}_{,A}\tilde{g}_{,B}\tilde{g}^{-1} \qquad (6.5)$$

$$\mu_{A\alpha} = e_\alpha (T^\alpha)_{AB} Z^B \tag{6.6}$$

In Eq. (6.5) we have introduced the modified superpotential

$$\tilde{g}(Z_A) \equiv Eg(Z_A); \; E \equiv \exp\left[\frac{\kappa^2}{4} \sum_A Z_A{}^2\right] \tag{6.7}$$

$\tilde{g}$ is a convenient variable since on the real manifold, $\tilde{g}_{,A} = EG_A$, where G_A is given by Eq. (3.2) with $Z^+ = Z$. Thus at the tree minimum, $\tilde{g}_{,A}$ have the sizes given by Eq. (4.6) and are thus very small (except perhaps for the super-Higgs channel). In addition to the above, there is, of course, the vector meson and gravitino mass matrices.

To illustrate the loop contribution to the effective potential we write down some of the terms for the part of Eq. (6.1) proportional to the super trace of M^4 [18]:

$$\begin{aligned} \mathrm{STr}M^4 = {} & [2(\tilde{g}_{,ABC}\tilde{g}_{,C})^2 + \ldots] + 2\kappa^2[(\tilde{g}_{,AB})^2(\tilde{g}_{,C})^2 \\ & - \tilde{g}_{,ABC}\tilde{g}_{,AB}\tilde{g}_{,C}\tilde{g} - 2\tilde{g}_{,ABC}\tilde{g}_{,A}\tilde{g}_{,B}\tilde{g}_{,C} + \ldots] \\ & + \kappa^4\left[-\frac{3}{2}\left|\tilde{g}_{,AB}\tilde{g}_{,AB}\tilde{g}^2\right| + 8\,\tilde{g}_{,AB}\tilde{g}_{,A}\tilde{g}_{,B}\tilde{g} + \ldots\right] \\ & + \kappa^6\left[\frac{3}{2}\,\tilde{g}_{,AA}\tilde{g}^3 + \ldots\right] + \frac{\kappa^8}{8}(\delta_{AA} - 2)\tilde{g}^4 \end{aligned} \tag{6.8}$$

In obtaining Eq. (6.8) there are the usual global cancellations as well as additional $O(\kappa^2)$ cancellations. (These cancellations are crucial in maintaining the gauge hierarchy.)

In the light sector, L_α, the condition determining the VEVs to one loop order is

$$\frac{\partial V^{tree}}{\partial L_\alpha} + \frac{\partial V^{(1)}}{\partial L_\alpha} = 0 \qquad (6.9)$$

For theories with a tree hierarchy and hence obeying Eqs. (4.3), the tree parts of Eq. (6.9) are $O(m_s{}^3)$ and hence the loop parts must be no larger if the gauge hierarchy is to be maintained. Using Eqs. (4.3), (4.6) and the fact that $g \sim O(m^2/\kappa)$, one sees that the only dangerous term in Eq. (6.9) is the boxed one whose derivative is of size

$$\kappa^4 \tilde{g}_{,AB}\tilde{g}_{,AB\alpha}\tilde{g}^2 \sim \tilde{g}_{,ij}\tilde{g}_{,ij\alpha}m_s{}^2 \qquad (6.10)$$

where the largest contribution occurs when A,B are in the heavy sector. A priori $\tilde{g}_{,ij\alpha} \sim O(1)$ and $g_{,ij}$ is of the GUT mass size M (since $\tilde{g}_{,ij}$ is proportional to the heavy field mass matrix). Thus in order to maintain the gauge hierarchy at the one loop level one requires the constraint on the superpotential, that at the tree minimum

$$\tilde{g}_{,ij}\tilde{g}_{,ij\alpha} \sim O(m_s) \qquad (6.11)$$

As an example, one notes that if one had a single coupling of the form

$$\lambda H_i H_j L_\alpha \qquad (6.12)$$

in g(Z) it would violate the gauge hierarchy at the loop level even though $\langle H_i \rangle = 0$ and one still had a tree hierarchy [19]. The one loop condition obtained above for maintaining the gauge hierarchy is Eq. (6.11), however, where the sum over i and j can cause cancellation of the dangerous $O(Mm_s^2)$ term in Eq. (6.10) [20].

Conditions for maintaining the gauge hierarchy up to the two loop level for a class of models possessing a "missing partner" scenario is given in Ref. [21].

VII. LOW MASS SECTOR IN SUPERGRAVITY GUT MODEL

One of the interesting aspects of Supergravity GUT models is that they make low energy predictions, some of which may be experimentally accessible. As an example we consider a simple model whose low mass sector $\{L_\alpha\}$ contains two Higgs doublets, H'_α and H^α ($\alpha = 4,5$) and a Higgs singlet U [22]. Thus there are five low energy chiral multiplets, plus 12 Majorana gaugino partners of the SU(3)xSU(2)xU(1) gauge mesons, as well as the usual three families of quark/lepton chiral multiplets [each containing their scalar superpartners (squarks and sleptons)]. To find the physical states, one must diagonalize the mass matrices that arise from the spontaneous symmetry breaking. In models where spontaneous breaking of SU(2)xU(1) occurs at the tree level (as in Sec. III) one generally has $\langle H'_5 \rangle = \langle H^5 \rangle$ to minimize the D part of Eq. (3.1). Those models which assume a heavy t-quark to trigger SU(2)xU(1) breaking through renormalization group effects [11] generally have $\langle H'_5 \rangle \ll \langle H^5 \rangle$. We summarize here results only for the first class of models [23] which have been discussed in Refs. [8], [9], [13], and [24].

(1) Higgs Spin 0 Mesons

The scalar components of the Higgs multiplets H'_α, H^α and U possess 5 complex fields or 10 real fields. These decompose into

(i) 3 massless Goldstone bosons (absorbed by the $W^{\pm}$ and Z^0 mesons); (ii) 1 charged scalar meson of mass $(m_H^{+})^2 = \bar{m}_H^2 + M_W^2$ where M_W is the $W^{\pm}$ mass and $\bar{m}_H$ is proportional to the gravitino mass m_g but of size that is model dependent; (iii) 5 neutral scalar mesons, one with mass $(m_H^{o})^2 = \bar{m}_H^2 + M_Z^2$ where M_Z is the Z^0 mass, and 4 others with mass of $O(m_g)$.

(2) Spin 1/2 Higgsinos and Gauginos

There are 5 L.H. Higgsino fields $\tilde{H}_{\alpha}'$, $\tilde{H}^{\alpha}$ (α - 4,5) and $\tilde{U}$, the four gauginos λ^i and λ of SU(2) and U(1) and the 8 gluinos λ^r (r = 1 . . . 8) of SU(3). At the tree level the gluinos ($\tilde{g}$) and the photino ($\tilde{\gamma}$) linear combination,

$$\lambda^{\gamma} = \cos\theta_W \lambda + \sin\theta_W \lambda^3 \tag{7.1}$$

remain massless, a result which is a direct consequence of Eq. (2.11a). (Here θ_W is the weak mixing angle.) However, the photino and gluino can grow masses at the one loop level through the interactions with the heavy GUT fields [e.g., the 24 representation $\sum_y^x$ in the model of Eq. (3.10)] required by the supersymmetrized Yang-Mills interactions [25]. One finds for the photino and gluino masses the result

$$m_{\tilde{\gamma}} = \frac{8}{3}\left(\frac{\alpha}{4\pi}\right)\bar{C}\, m_g \sim (1 - 5)\text{GeV} \tag{7.2a}$$

$$m_{\tilde{g}} = \frac{\alpha_s}{4\pi}\bar{C}\, m_g \sim (5 - 25)\text{GeV} \tag{7.2b}$$

where $\bar{C}$ is proportional to the Casimir operators of the heavy fields, and α and α_s are the QED and QCD coupling constants respectively.

The existance of light photinos and gluinos would be strong evidence for the validity of Eq. (2.11a).

The remaining 8 gauginos and Higgsinos mix and form massive particles. As pointed out by Weinberg [8], Eq. (2.11a) implies in general that there is always one charged fermion (called a Wino) lying below the W meson and one Majorana fermion (called a Zino) lying below the Z meson. For the model under consideration, there are actually two charged Winos, $\tilde{W}_{(+)}$ and $\tilde{W}_{(-)}$ with masses $\tilde{m}_+$ and $\tilde{m}_-$ obeying

$$\tilde{m}_+\tilde{m}_- = M_W^2 \tag{7.3}$$

and two Majorana Zinos, $\tilde{Z}_+$ and $\tilde{Z}_-$, with masses $\tilde{\mu}_+$ and $\tilde{\mu}_-$ obeying

$$\tilde{\mu}_+\tilde{\mu}_- = M_Z^2 \tag{7.4}$$

In addition, the following mass splitting and mass sum rule (involving Wino, W and Higgs mesons) hold

$$\tilde{m}_+ - \tilde{m}_- = \tilde{\mu}_+ - \tilde{\mu}_- \tag{7.5}$$

$$m_g^2 + [M_W^2 - \tilde{m}_-^2]^2/\tilde{m}_-^2 = \tfrac{1}{2}[(m_H^+)^2 - M_W^2] \tag{7.6}$$

Finally, there are two additional Majorana Higgsinos whose masses are model dependent and $O(m_g)$ [26].

(3) Squarks and Sleptons

Associated with Weyl spinor is a complex scalar. Hence each massive quark (u, d, . . .) and lepton (e, μ, . . .) have two complex scalar superpartners which we label $\tilde{u}_{\pm}$, $\tilde{d}_{\pm}$, . . . and $\tilde{e}_{\pm}$, $\tilde{\mu}_{\pm}$, ... The massless neutrinos each have a single superpartner, the sneutrinos $\tilde{\nu}_e$, $\tilde{\nu}_\mu$, ... These particles have masses close to the gravitino mass and are nearly degenerate with each other. Thus one finds

$$m_{\tilde{\nu}} = m_g \tag{7.7}$$

$$m_{\tilde{e}_{\pm}}{}^2 = m_{\tilde{\nu}}{}^2 + m_e{}^2 \pm \beta m_e m_{\tilde{\nu}} \tag{7.8}$$

$$m_{\tilde{u}_{\pm}}{}^2 = m_{\tilde{\nu}}{}^2 + m_u{}^2 \pm \beta m_u m_{\tilde{\nu}}, \text{ etc.} \tag{7.9}$$

where $\beta \sim O(1)$ is a model dependent factor. The fact that the squarks come out automatically to be nearly degenerate naturally suppresses flavor changing neutral currents.

Once the field combinations which diagonalize the mass matrices are determined, one may calculate the three point vertices between these physical fields by eliminating the original fields in terms of them. The basic vertices of interest for phenomenological applications are the following:

$$W - \tilde{W} - \tilde{\gamma};\ W - \tilde{W} - \tilde{Z};\ Z - \tilde{W} - \tilde{W} \tag{7.10}$$

$$\tilde{W} - q(\ell) - \tilde{q}_{\pm}(\tilde{\ell}_{\pm});\ \tilde{Z} - q(\ell) - \tilde{q}_{\pm}(\ell_{\pm}) \tag{7.11}$$

and

$$\tilde{q}_{\pm} - q - \tilde{g}(\tilde{\gamma});\ \tilde{\ell}_{\pm} - \ell - \tilde{\gamma} \tag{7.12}$$

where q = quark, $\tilde{q}_{\pm}$ = squark, ℓ = lepton, etc. Eq. (7.10) governs decay modes of the $W^{\pm}$ and Z^0 mesons into gauginos (Winos, Zinos and photinos), while Eqs. (7.11) and (7.12) are the vertices that enter into Wino and Zino decays. Each of these vertices is completely determined in terms of the masses of the particles and the weak angle θ_W. For example, the interaction between the low lying Zino, the up quark u and up squarks $\tilde{u}_{\pm}$ is

$$L_{\tilde{Z}u\tilde{u}} = i \frac{e}{\sqrt{2}} \frac{M_Z^2}{\tilde{\mu}_-^2 + M_Z^2} [- \frac{1}{4} (\frac{5}{3} \tan\theta_W - \cot\theta_W)\bar{u}(\gamma_5\tilde{u}_+ - \tilde{u}_-)$$

$$+ (2sm\theta_W)^{-1}\bar{u}(\gamma_5\tilde{u}_- - \tilde{u}_+)]\tilde{Z}_{(-)} \qquad (7.13)$$

and so forth for the other vertices.

VIII. SUPERSYMMETRIC DECAY OF $W^{\pm}$ AND Z^0 MESONS

The vertices of Eq. (7.10) imply the possibility of new supersymmetric decay modes of the $W^{\pm}$ and Z^0 mesons provided the gauginos are sufficiently light. Thus one can have

$$W^{\pm} \rightarrow \tilde{W}_{(-)}^{\pm} + \tilde{\gamma};\quad \tilde{m}_{(-)} + m_{\tilde{\gamma}} < M_W \qquad (8.1)$$

$$W^{\pm} \rightarrow \tilde{W}_{(-)}^{\pm} + \tilde{Z}_{(-)};\quad \tilde{m}_{(-)} + \tilde{\mu}_{(-)} < M_W \qquad (8.2)$$

$$Z \rightarrow \tilde{W}_{(-)}^{+} + \tilde{W}_{(-)}^{-};\ 2\tilde{m}_{(-)} < M_Z \qquad (8.3)$$

Since the low lying Wino, $\tilde{W}_{(-)}$, generally lies below the W meson for all models obeying Eq. (2.11a) [8], decay (8.1) will almost always be energetically possible provided the photino does not become too heavy, which is also expected from Eq. (2.11a). [See e.g. Eq. (7.2a).] It is less clear theoretically whether decays of Eqs. (8.2) and (8.3) are energetically possible. However, the mass relations of Eqs. (7.3) - (7.5) show there is a wide band of allowed Wino and Zino masses that allow these decays.

As a characteristic example, we assume a Wino of mass $\tilde{m}_{(-)} = 30$ GeV and hence by Eqs. (7.3) - (7.5) a Zino of mass $\tilde{\mu}_{(-)} = 37$ GeV. The branching ratios for Eqs. (8.1) - (8.3) then are [8], [24], [27]

$$W^{\pm} \rightarrow \tilde{W}^{\pm} + \tilde{\gamma};\ 4.0\% \tag{8.4}$$

$$W^{\pm} \rightarrow \tilde{W}^{\pm} + \tilde{Z};\ 15\% \tag{8.5}$$

$$Z^{0} \rightarrow \tilde{W}^{+} + \tilde{W}^{-};\ 19\% \tag{8.6}$$

For comparison we note that for the standard decay mode, $W \rightarrow e + \nu_e$, the branching ratio is about 6.8%, while the $Z \rightarrow e^+ + e^-$ branching ratio is about 2.5%. Thus the supersymmetric decay rates are quite large. In order to see how to recognize the existance of these new decay channels, it is necessary to investigate what the Wino and Zino decay into. The dominant decay processes are shown in Fig. 1. We consider here for definiteness the case where the squark mass $m_{\tilde{q}}$ equals the W mass. Then (1a) and (1b) are comparable in size and (1c) and (1d) are comparable in size. One would expect diagrams (1a) and (1c) to yield a 3 jet final state (provided the

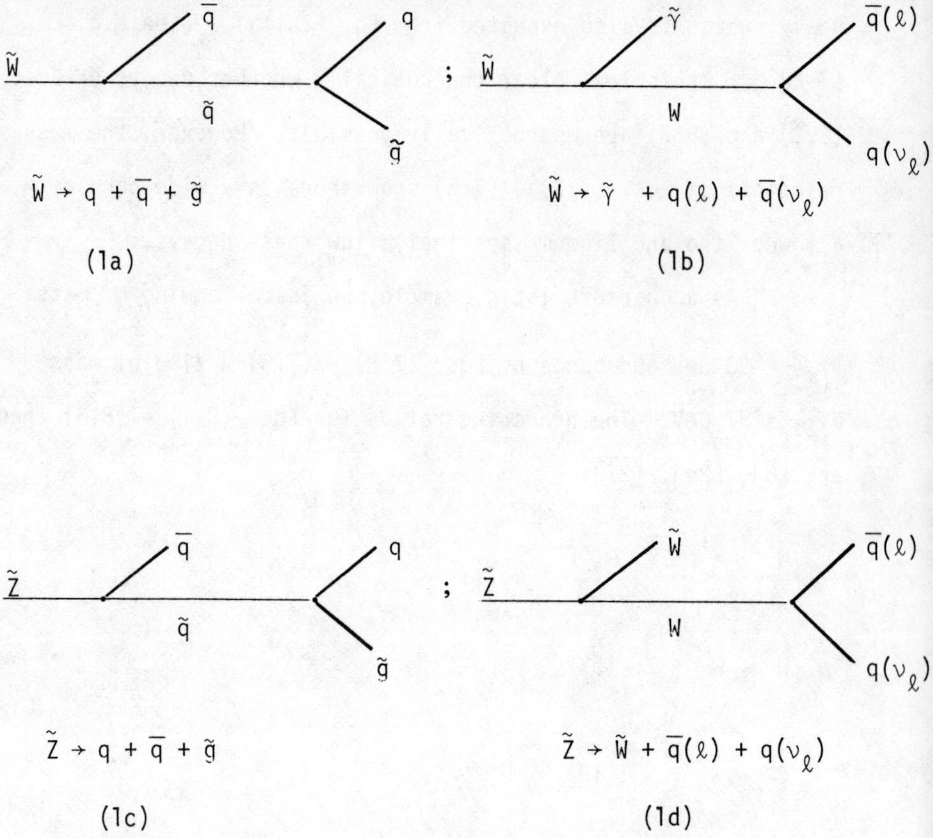

Fig. 1. Dominant decay modes of the Wino and Zino. If the squark masses, $m_{\tilde{q}}$, are comparable to the W mass, then (1a) $\approx$ (1b) and (1c) $\approx$ (1d). For $m_{\tilde{q}} \gtrsim 3M_W$, (1b) and (1d) dominate. (In the diagrams q = quark, $\tilde{q}$ = squark, ℓ = lepton, $\tilde{\gamma}$ = photino.)

gluino is not too heavy) while the final state of (1b) would be $\tilde{\gamma}$ + 2 jets (and the leptonic mode $\tilde{\gamma} + \ell + \nu_\ell$). Diagram (1d) yields a Wino (which then decays via (1a) or (1b)) plus hadrons (or leptons).

The branching ratios for the various final states can be calculated using the vertices of Eqs. (7.11) and (7.12). The rates for some of the channels that are most likely to be experimentally accessible are given in Table 1. (Phase space corrections have only been evaluated approximately.) Each supersymmetric decay mode must of course compete against various backgrounds, but each has a characteristic signal that might possibly make them observable. Thus in processes (1) and (2) the jets would show large unbalanced momentum. In process (3) the charged lepton and hadrons should be in opposite hemispheres. Thus such signals might be observable now at CERN. Processes (4) and (5) have low branching ratios but have rather different final states than in standard W decay modes and so might be observable in a future high statistic experiment. (Again in (4), the two leptons from the Zino decay should be largely in opposite hemispheres from the hadrons.)

The supersymmetric Z decay modes (6) and (7) of Table 1 are remarkably large and again have a signal distinct from conventional decay modes of the Z in that only one lepton appears. (This lepton again should be opposite to the hadrons.) One might expect 100 of such event/day at SLC, and so such modes may be observable at SLC or LEP.

Table 1. Preliminary calculations of relative branching ratios of various supersymmetric decays of the $W^{\pm}$ and Z^{o} mesons. The $W^{\pm}$ branching ratios are relative to $W^{+} \to e^{+} + \nu_{e}$ and the Z^{o} branching ratios are relative to $Z^{o} \to e^{+} + e^{-}$. (ℓ, ℓ_1, ℓ_2 = $e^{\pm}$ or $\mu^{\pm}$) The calculations are for a Wino of mass 30 GeV and for the squark mass equal to the W mass.

Process	Relative Branching Ratio (%)	Mode
1) $W \to \tilde{\gamma} + (3jets + \tilde{\gamma})$	25	$W \to \tilde{\gamma} + \tilde{W}$
2) $W \to \tilde{\gamma} + (2jets + \tilde{\gamma})$	25	$W \to \tilde{\gamma} + \tilde{W}$
3) $W \to (\ell + \nu_{\ell} + \tilde{\gamma}) + (hadrons)$	30	$W \to \tilde{W} + \tilde{Z}$
4) $W \to (hadrons + \tilde{\gamma}) + (\ell_1 + \ell_2 + \nu_1 + \nu_2 + \tilde{\gamma})$	$\lesssim 1$	$W \to \tilde{W} + \tilde{Z}$
5) $W \to \ell_1 + \ell_2 + \ell_3 + \nu_1 + \nu_2 + \nu_3 + 2\tilde{\gamma}$	$\lesssim 1$	$W \to \tilde{W} + \tilde{Z}$
6) $Z \to (\ell_1 + \nu + \tilde{\gamma}) + (3jets + \tilde{\gamma})$	95	$Z \to \tilde{W} + \tilde{W}$
7) $Z \to (\ell_1 + \nu + \tilde{\gamma}) + (2jets + \tilde{\gamma})$	95	$Z \to \tilde{W} + \tilde{W}$

Finally, we mention that if the photino is as light as indicated by Eq. (7.2a), it may be pair produced at PEP or PETRA.

In summary, if gauginos are as light as this class of Supergravity GUT models suggest, then there is a reasonable possibility that experimental tests of these ideas are feasible.

ACKNOWLEDGEMENTS

We are pleased to acknowledge conversations with C.S. Aulakh, L. Hall, J. Polchinski, D. Shambroom, E. von Goeler, and S. Weinberg.

FOOTNOTES

1. K.S. Stelle and P.C. West, Phys. Lett. 74B 330 (1978); S. Ferrara and P. van Nieuwenhuizen, Phys. Lett. 74B 333 (1978).
2. S. Ferrara and P. van Nieuwenhuizen, Phys. Lett. 76B 404 (1978); K.S. Stelle and P.C. West, Phys. Lett. 77B 376 (1978).
3. E. Cremmer, B. Julia, J. Scherk, S. Ferrara, L. Girardello and P. van Nieuwenhuizen, Nucl. Phys. B147 105 (1979).
4. A.H. Chamseddine, R. Arnowitt and P. Nath, Phys. Rev. Letters 49 970 (1982).
5. E. Cremmer, S. Ferrara, L. Girardello, and A. van Proeyen, Phys. Lett. 116B 231 (1982); CERN preprint TH. 3348 (1982).
6. J. Bagger and E. Witten, Phys. Lett. 118B 103 (1982); J. Bagger, Nucl. Phys. B211 302 (1983).
7. S. Ferrara, J. Scherk, and P. van Nieuwenhuizen, Phys. Rev. Letters 37 1035 (1976).
8. S. Weinberg, Phys. Rev. Letters 50 387 (1983).

9. L. Hall, J. Lykken and S. Weinberg, University of Texas preprint UTTG-1 (1983).

10. J. Polony, Budapest Preprint KFKI-1977-93 (1977) (unpublished).

11. Use of the renormalization group to discuss radiative breaking of SU(2)xU(1) have been made by J. Ellis, D.V. Nanopoulos and K. Tamvakis, Phys. Lett. 121B 123 (1983); L.E. Ibanez, Univsersidad Autonoma de Madrid preprint, FTUAM/82-8 (1982); L. Alvarez-Gaume, J. Polchinski, and M.B. Wise, Harvard/CALT preprint HUTP-82/A063/ CALT-68-990 (1982); J. Ellis, J.S. Hagelin, D.V. Nanopoulos and K. Tamvakis, SLAC preprint SLAC-PUB-3042 (1983). This approach generally requires a heavy t-quark to accomplish the SU(2)xU(1) breaking ($m_t \gtrsim 100$ GeV). For a discussion of these results see the talks by J. Ellis and L. Ibañez at this workshop.

12. Eq. (3.10) is the superpotential of the global SUSY models of S. Dimopoulos and H. Georgi, Nucl. Phys. B193 150 (1981) and N. Sakai, Z. Phys. C11 153 (1981) with the addition of the singlet term.

13. P. Nath, A.H. Chamseddine, and R. Arnowitt, Harvard/Northeastern preprint HUTP-82/A057/NUB#2579 (1982).

14. More precisely, the interaction terms between Z and Z_a in $g(Z_a,Z)$ must be quite small for the protection theorem to hold. For example, there could be couplings in g of the form $\lambda_{ij}H_iH_jZ$ or $\lambda_{\alpha\beta}H_\alpha H_\beta Z$ provided $\lambda_{ij}, \lambda_{\alpha\beta} \sim O(\kappa m_s)$. A term of the form $\lambda_i H_i Z^2$ requires $\lambda_i \sim O(\kappa^2 m_s M)$ etc.

15. Previous discussions by R. Barbieri, S. Ferrara and C.A. Savoy, Phys. Lett. 119B 343 (1983) and H.P. Nilles, M. Srednicki and D. Wyler, have considered only a partial low energy effective

16. More precisely, in writing down Eq. (3.4), we restricted Z_A to the real manifold. To obtain U, one must generalize Eqs. (4.9)-(4.11) to complex fields (which is a straight forward extension) and then carry out the indicated elimination.

17. A more detailed discussion of the properties of the low energy effective potential is given in Refs. [9] and [13] and in the talk by S. Weinberg at this workshop.

18. R. Arnowitt, A.H. Chamseddine and P. Nath, Phys. Lett. 120B 145 (1983).

19. The fact that a single coupling of the form Eq. (6.12) violates the gauge hierarchy at the loop level was discussed in [18], and more recently by H.P. Nilles, M. Srednicki and D. Wyler, CERN preprint TH. 3461 (1982) and B. Lahanas, CERN preprint TH. 3467 (1982) within the framework of special models. The latter authors have pointed out that because of this, the gauge hierarchy of the model of Eq. (3.10) is destabilized by loop corrections due to the Higgs triplet H_3 couplings $\lambda_3 H_{3i}' H_3{}^i U$ (even though $\langle H_3{}^i\rangle = 0$ due to color conservation).

20. For example, one may easily modify the model of Eq. (3.10) so that the one loop gauge hierarchy is maintained by introducing an additional pair of Higgs multiplets, K_x' and K^x and replacing $\lambda_3 U H_x' H^x$ by $U(\lambda_3 H_x' K^x + \lambda_3' K_x' H^x)$ and adding $\lambda_2' K'(\Sigma + 3M)K$ to $g_1(Z)$. The Higgs triplets are superheavy and Eq. (6.11) is automatically satisfied with α in the light U channel. We note also that the $\ln M^2$ factor in Eq. (6.1) implies additional one loop constraints which are also satisfied in the above modification of the model of Eq. (3.10).

21. S. Ferrara, D.V. Nanopoulos and C.A. Savoy, CERN preprint TH. 3442 (1982).

22. A supersymmetry GUT model requires at least two Higgs doublets to account for both up and down quark masses.

23. Some of the phenomenological consequences of models with $\langle H_5' \rangle \ll \langle H^5 \rangle$ can be found in [11]. An analysis of a general class of models with arbitrary ratio $\langle H_5' \rangle / \langle H^5 \rangle$ is in preparation, and will be presented elsewhere.

24. R. Arnowitt, A.H. Chamseddine and P. Nath, Phys. Rev. Letters 50 232 (1983).

25. R. Arnowitt, A.H. Chamseddine and P. Nath (unpublished); L. Alvarez-Gaumé, J. Polchinski and M.B. Wise, Ref. [11].

26. The Wino and Zino mass relations are a direct consequence of the condition $\langle H_5' \rangle = \langle H^5 \rangle$. Thus for arbitrary Higgs VEVs, Eq. (7.3) becomes $\tilde{m}_+\tilde{m}_- = |\sin 2\alpha| M_W^2$ where $\tan\alpha \equiv \langle H_5' \rangle / \langle H^5 \rangle$. More complicated generalizations hold in the Zino sector also,

27. P. Nath, R. Arnowitt and A.H. Chamseddine, "Wino and Zino Decay of the W and Z Meson," Northeastern Univ. preprint NUB#2588.

THE MONOPOLE CATALYSIS S-MATRIX

CURTIS G. CALLAN, JR.
JOSEPH HENRY LABORATORIES
PRINCETON UNIVERSITY
PRINCETON NJ 08540

ABSTRACT

We discuss the problem of computing the S-matrix for scattering quarks and leptons from magnetic monopoles. We show that it is possible to make estimates of the relative strength of baryon-number-violating and baryon-number-conserving inclusive cross-sections. We pay particular attention to the role of fermion masses and explore the way in which heavy fermion degrees of freedom "freeze out" of low-energy scattering processes.

INTRODUCTION

The goal of any theoretical discussion of monopole-catalysed proton decay is the quantitative calculation of the S-matrix for hadron-monopole scattering. To date, only the more limited goal of showing that proton decay catalysis cross-sections are of "typical strong-interaction magnitude" has been achieved [1,2]. In this talk, we will show how to use the field theory formalism which has been developed for the study of this problem to extract much more detailed information about the S-matrix. Our main result is that, while cross -sections to individual final state channels are impossible to estimate, inclusive cross sections, to baryon-number-violating final states for example, can in some circumstances be reliably estimated.

Two subsidiary, but important, matters which we are able to clear up in the course of this discussion, are the effect of varying fermion masses and numbers of generations on the catalysis cross section, as well as the detailed manner in which conservation laws (color, charge and weak isospin) are implemented. These results enable us to dispose of two worries which have been expressed concerning proton decay catalysis: that the baryon-number-violating cross section could be suppressed by the existence of many massive flavors or by an electroweak barrier due to weak isospin Coulomb energy effects [3].

THE MODEL

To keep the problem manageable and to focus on essentials, we will ignore confinement effects and study the scattering of quarks and leptons from a monopole. Since we are mainly interested in inclusive cross section questions, this should not lead us too far astray. The field theory model we have developed in our earlier studies of monopole-fermion interactions is well-suited to this problem and we shall make heavy use of it. Although for completeness, we shall give a description of the essentials of the model, we assume the reader to be familiar with its derivation, especially in the form given in my talk to the Wingspread Conference on Magnetic Monopoles [4].

We make the usual assumptions of studying the lightest SU_5 monopole, keeping only the s-wave fermion degrees of freedom and using the bosonization trick to convert the system to an equivalent one-dimensional scalar theory. The scalar fields we need are $\varphi_{ai}(r)$, where a=1,2,3 is the generation index and $i=e^+,\overline{d}_3,u_1,u_2$ labels the fermion species which actually have an s-wave. The scalar Lagrangian describing free propagation in the background field of the monopole is

$$L_\varphi=\int_0^\infty dr\sum_{ai}\left[\frac{1}{2}\left(\partial_t\varphi_{ai}\right)^2-\frac{1}{2}\left(\partial_r\varphi_{ai}\right)^2+m_{ai}\mu\cos 2\sqrt{\pi}\varphi_{ai}\right] \tag{2.1}$$

where m_{ai} are the fermion bare masses and μ is a renormalization mass parameter. This is a sum of Sine-Gordon theories and the Sine-Gordon soliton (interpolating between $N\sqrt{\pi}$ and $(N+1)\sqrt{\pi}$) is the scalar theory realization of the underlying fermion.

All of the physics of the interaction of the fermions with the monopole core is contained in the boundary conditions to be imposed at r=0. If we were to impose free boundary conditions, an inward moving soliton would pass through r=0 (into the monopole core) and vanish without a trace! A careful study of what goes on when s-wave fermions interact with the the core shows that the right behavior is implied by the following mixture of free and fixed boundary conditions:

$$\varphi_{ae^+}(0)=\varphi_{a\overline{d}_3}(0) \qquad \varphi_{au_1}(0)=\varphi_{au_2}(0) \qquad a=1,2,3 \tag{2.2.a}$$

$$\sum_{ai}\varphi_{ai}(0)=0 \tag{2.2.b}$$

$$\sum_a A_a^r\left[\varphi'_{ae^+}(0)+\varphi'_{a\overline{d}_3}(0)\right]+\sum_a B_a^r\left[\varphi'_{au_1}(0)+\varphi'_{au_2}(0)\right]=0 \qquad r=1,\dots,5 \tag{2.2.c}$$

where the coefficients A_a^r and B_a^r are the five independent solutions of

$$\sum_{a=1}^{3}\left[A_a^r+B_a^r\right]=0 \tag{2.2.d}$$

These conditions look complicated, but have simple physical interpretations. The six conditions of (2.2.a) express the fact that e^+ and $\overline{d}_3$ (u_1 and u_2) fermions are not really independent fields: an ingoing $e^+(u_1)$ is automatically converted by the SU_5 monopole core fields into an outgoing $d_3(\overline{u}_2)$. This sort of event deposits charge on the monopole core and excites the dyon degree of freedom. Since the minimum electrostatic energy associated with dyon excitation is of order $e^2M_X\approx 10^{12}Gev$, we want to impose the condition that under the joint action of all the fields, the dyon is never excited. That is the function of the condition of (2.2.b). The remaining 5 conditions are that the field combinations orthogonal to the previous 7 satisfy free boundary conditions. The total of 12 conditions corresponds to the total number of s-wave fermions in 3 generations. If we had only one generation, the conditions would take on the simpler form

$$\varphi_{e^+}(0)=\varphi_{\overline{d}_3}(0) \qquad \varphi_{u_1}(0)=\varphi_{u_2}(0) \tag{2.3.a}$$

$$\varphi_{e^+}(0)+\varphi_{\overline{d}_3}(0)+\varphi_{u_1}(0)+\varphi_{u_2}(0)=0 \tag{2.3.b}$$

$$\varphi'_{e^+}(0)+\varphi'_{\overline{d}_3}(0)-\varphi'_{u_1}(0)-\varphi'_{u_2}(0)=0 \tag{2.3.c}$$

One of our goals is to understand how (2.3) reduces to (2.2) as the masses of the second and third generations increase to infinity.

CHARGE CONSERVATION

The boundary conditions have been designed to veto excitation of the dyon degree of freedom of the monopole core. Since all of the charges of $SU_3^c x SU_2^w x U_1^{em}$ carry Coulomb energy we might worry that a veto is required for each of them individually, while only one condition has been devoted to this task. The question is whether the boundary conditions of (2.2) actually make the fluxes of all the relevant charges into the point r=0 (i.e. onto the monopole core) vanish. For a charge $\hat{Q}$, the question is whether, as an automatic consequence of the boundary conditions

$$4\pi r^2 J_r^{\hat{Q}}\Big|_{r=0} = 0 \qquad (3.1)$$

where J_r^Q is the radial component of the $\hat{Q}$ charge current.

According to the general discussions of bosonization, the currents of Abelian charges are simple linear combinations of space and time derivatives of fields. For the case of one fermion generation, we can easily write out the radial charge currents of the diagonal charges of $SU_3^c x SU_2^w x U_1^{em}$:

$$4\pi r^2 J_r^Q = \frac{1}{\sqrt{\pi}}\left[\dot{\varphi}_{e^+} + \frac{1}{3}\dot{\varphi}_{\bar{d}_3} + \frac{2}{3}\dot{\varphi}_{u_1} + \frac{2}{3}\dot{\varphi}_{u_2}\right] \qquad (3.2.a)$$

$$4\pi r^2 J_r^{Y_c} = \frac{1}{\sqrt{\pi}}\left[\frac{2}{3}\dot{\varphi}_{\bar{d}_3} + \frac{1}{3}\dot{\varphi}_{u_1} + \frac{1}{3}\dot{\varphi}_{u_2}\right] \qquad (3.2.b)$$

$$4\pi r^2 J_r^{Y_{3c}} = \frac{1}{\sqrt{\pi}}\left[\frac{1}{2}\dot{\varphi}_{u_1} - \frac{1}{2}\dot{\varphi}_{u_2}\right] \qquad (3.2.c)$$

$$4\pi r^2 J_r^{I_3^w} = \frac{1}{4\sqrt{\pi}}\left[\varphi'_{e^+} + \varphi'_{\bar{d}_3} - \varphi'_{u_1} - \varphi'_{u_2} - \dot{\varphi}_{e^+} - \dot{\varphi}_{\bar{d}_3} - \dot{\varphi}_{u_1} - \dot{\varphi}_{u_2}\right] \qquad (3.2.c)$$

If we consult (2.3), we immediately see that the boundary conditions cause all of these fluxes to vanish. Exactly the same thing happens for the multigeneration case. The question of conservation of the off-diagonal, non-Abelian charges of SU_3 is not so easy to discuss in the bosonized language and we shall say no more about it here.

The conclusion of this discussion is that all relevant gauge charges are conserved at short distances (i.e. on the scale of the monopole core) and no Coulomb energy barrier associated with leakage of charge onto the monopole core can arise. Weak isospin is of course not conserved as a general proposition, but its non-conservation will be due to "soft" physics such as mass terms.

FREEZE-OUT OF HEAVY FERMIONS

If we want to study scattering at center of mass energy E, it is surely not necessary to include fermions whose mass is much greater than E. We would now like to derive the recipe for eliminating heavy degrees of freedom from the scalar field Lagrangian and the boundary conditions.

Let $m_h \gg E$ be the mass of a typical heavy degree of freedom φ_h. Because of the associated mass term, φ_h is constrained to be nearly zero everywhere. On the other hand, through the boundary conditions, φ_h is driven at r=0 by the variation of the light degrees of freedom. Therefore a more accurate picture of the space and time dependence of φ_h is

$$\varphi_h = Ae^{-m_h r} e^{-iEt} \qquad (4.1)$$

The spatial scale of the falloff is of course set by m_h, the temporal scale is set by the energy of the scattering process and the normalization, A, has to be determined. To determine A we note that some of the boundary conditions (2.2.c) have the form

$$\varphi'_h(0)=\sum\varphi'_{light}(0) \tag{4.2}$$

Since the typical spatial derivative of the light fields is of order E, this means that $\varphi'_h(0)$ is of order E also. Therefore A must be of order E/m_h and

$$\varphi_h \approx \frac{E}{m_h} e^{-m_h r} e^{-iEt} \tag{4.3.a}$$

$$\varphi'_h \approx E e^{-m_h r} e^{-iEt} \tag{4.3.b}$$

In other words, if $E<<m_h$, φ_h and its derivatives are negligible everywhere, except that φ'_h is of order E (the same size as φ'_{light}) in a region of size m_h^{-1} around the origin.

The dynamics of the remaining light degrees of freedom are therefore obtained by a) omitting φ_h from the Lagrangian b) setting $\varphi_h(0)=0$ in the boundary conditions which involve only the values of the fields c) dropping from consideration any boundary condition involving the derivatives of the fields which contains φ'_h. The reason for c) is that since $\varphi'_h(0)$ is just as big as φ'_{light}, such boundary conditions do not provide a constraint on the light fields and are not part of the dynamical system involving the light fields only.

What happens to the boundary conditions of (2.2) if we make one or more whole generation of fermions infinitely massive? According to the above rules, one simply lets the generation index, a, wherever it appears, run only over the light generations. Having done that, the number of independent solutions of (2.2.d) and the number of independent derivative boundary conditions will depend on the number of surviving generations. For the case of one surviving generation this recipe reproduces the conditions of (2.3). In general there are as many conditions as fields.

In the real world, low energy physics brings into play "one-and-a-half" generations: the first generation plus the $\mu^+\bar{s}_3$ pair of the second generation. If we freeze out the remaining fields, the boundary conditions on the six remaining fields turn out to be

$$\varphi_{e^+}(0)=\varphi_{\bar{d}_3}(0) \quad \varphi_{u_1}(0)=\varphi_{u_2}(0) \quad \varphi_{\mu^+}(0)=\varphi_{\bar{s}_3}(0) \tag{4.4.a}$$

$$\varphi_{e^+}(0)+\varphi_{\bar{d}_3}(0)+\varphi_{u_1}(0)+\varphi_{u_2}(0)+\varphi_{\mu^+}(0)+\varphi_{\bar{s}_3}(0)=0 \tag{4.4.b}$$

$$\varphi'_{e^+}(0)+\varphi'_{\bar{d}_3}(0)-\varphi'_{u_1}(0)-\varphi'_{u_2}(0)=0 \tag{4.4.c}$$

$$\varphi'_{e^+}(0)+\varphi'_{\bar{d}_3}(0)+\varphi'_{u_1}(0)+\varphi'_{u_2}(0)-2\varphi'_{\mu^+}(0)-2\varphi'_{\bar{s}_3}(0)=0 \tag{4.4.d}$$

These boundary conditions are of course relevant only for discussing scattering at energies significantly below the charm quark mass. As we increase that energy past the various mass thresholds we are obliged to resurrect the corresponding fields.

An interesting point about the boundary conditions (4.4) concerns the conservation of charges. If we carry out an analysis similar to that of Sect. 3, but including only the surviving light fields in the currents we will find that although the flux of color and ordinary charge into the origin still vanishes, the flux of weak isospin does not! This is because the full weak isospin current (3.2.c) contains derivatives of the heavy fields. According to (4.3), these terms are not negligible in a region of size m_h^{-1} about the origin and should not really have been dropped from the current. Although the heavy fields play no role in the dynamics, they are responsible, via their mass terms, for weak isospin violation in the light sector! In this case, of course, the breaking of weak isospin in the light sector is very explicit, since we have given a very different mass to the two members of the c-s doublet.

SCATTERING SOLUTIONS

Let us now discuss what happens when we scatter a fermion from the monopole. The general arguments of [1,2] suggest that baryon-number-violating processes such as

$$e^{+} + M \to M + u_1 + u_2 + d_3$$

will have large cross-section (of order π / E_e^2) but do not tell us about details such as how the S-matrix depends on the number of final state particles, on the net change in baryon number, etc. To answer such questions we have, in principle, only to scatter a fermion soliton from the monopole and see what comes out.

Despite the close connection of the system of (2.1.2) with the Sine-Gordon theory, it is not exactly soluble, even classically. However, if we consider scattering at center of mass energy large compared to the fermion masses, we may drop the $\cos 2\sqrt{\pi}\varphi$ mass terms. The system is then quadratic with linear boundary conditions and exactly soluble. For simplicity let us temporarily imagine that masses and energies are such that we may neglect the masses of the first generation fermions and regard the second and third generation fermions as frozen out.

For definiteness, let us consider the scattering of a positron soliton from the monopole. The initial condition, shown in Fig.1, is a φ_e^+ soliton

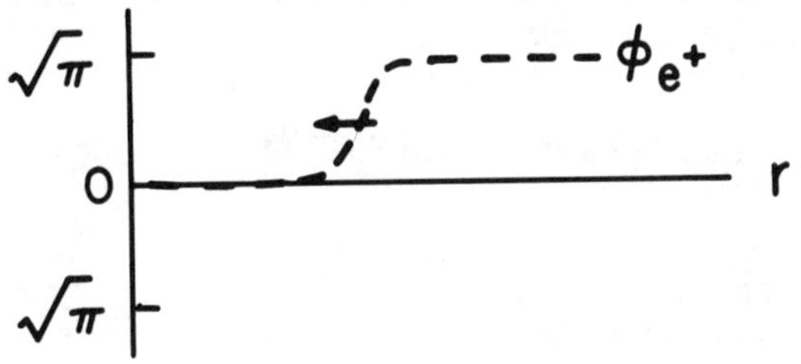

Figure 1. An incoming positron incident on the monopole

interpolating between 0 and $\sqrt{\pi}$ (to correspond to exactly one positron), moving toward the monopole. In the zero mass approximation, we have only to solve the massless Klein-Gordon equation subject to the boundary conditions of (2.3). This is a simple matter and we easily find the outgoing state shown in Fig. 2. This final state is a collection of outgoing half-solitons ,leaving behind them a displaced vacuum! In particle language we have

$$e^{+}_{L} + M \to M + \tfrac{1}{2}e^{+}_{R} + \tfrac{1}{2}u_{1_R} + \tfrac{1}{2}u_{2_R} + \tfrac{1}{2}d_{3_L}$$

Although strange, the answer is quite definite and we must decide how to interpret it [5].

First, note that, whatever else has happened, all the conservation laws which should be obeyed, are obeyed. Interpreting the charges of a half-soliton in the obvious way, it is easy to verify that the $SU_3 x SU_2 x U_1$ charges of the

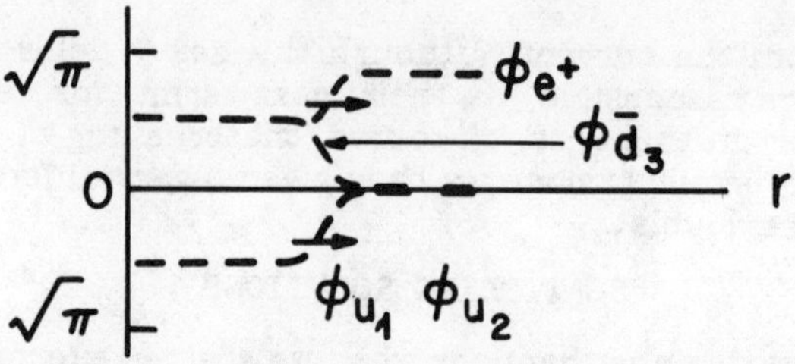

Figure 2. The outgoing state resulting from an incident positron.

ingoing and outgoing states are the same. On the other hand, both baryon number and lepton number change by one-half unit in such a way that B-L is conserved. The non-conservation of baryon number comes from the boundary conditions which in turn reflect the appearance of baryon-number-violating SU_5 fields inside the monopole.

The fractional-fermion outcome of the scattering actually has a sensible quantum-mechanical interpretation. Suppose the "out" state were

$$|out\rangle = \frac{1}{\sqrt{2}}|e^+{}_R\rangle + \frac{1}{\sqrt{2}}|u_{1_R} u_{2_R} d_{3_R}\rangle$$

The expectation in this state of baryon number, positron number, d_{3_L} number, etc. are all ½, just as in the scalar field configuration specified in Fig. 2. This suggests that the true out state for this process has a decomposition in terms of fermion Fock states of the form

$$|out\rangle = \sum_i a_i |e^+{}_R, i\rangle + \sum_i b_i |u_{1_R} u_{2_R} d_{3_L}, i\rangle$$

where $e^+{}_R, i$ indexes states with any number of fermions and antifermions but with the same net quantum numbers as e^+ or $u_1 u_2 d_3$ and where, in order to guarantee that the net baryon number of the state is one-half,

$$\sum_i |a_i|^2 = \sum_i |b_i|^2 = ½.$$

In terms of inclusive cross-sections, this amounts to the simple statement that the inclusive cross-section to go to final states with baryon number zero or one respectively, are each equal to one-half the total cross-section for $e^+ + M$ to go to anything! On conservation of probability grounds we know that the total s-wave cross-section is π/E^2 (where E is the incident electron energy), so we have the prediction that

$$\sigma(e^+M \to MX, \Delta B=0) = ½\frac{\pi}{E^2}$$

$$\sigma(e^+M \to MX, \Delta B=1) = ½\frac{\pi}{E^2}$$

The equality sign has to be taken with a grain of salt because we have neglected all "final state interactions" due to color forces, etc. All we really want to abstract from this is the conclusion that the fermion monopole S-matrix is in general a very complicated object, but with roughly a 50-50 distribution between $\Delta B=0$ and $\Delta B=1$ final states. The peculiar appearance of fractional solitons in the scattering solution is the only way the scalar field theory can realize this rather simple quantum mechanics.

A complementary view of these phenomena can be obtained by considering the role of mass terms. In the massless limit, the ground state has a continuous degeneracy: the energy is minimized by any spatially constant, time-independent set of fields satisfying the boundary conditions of (2.3) ;the most general solution is

$$\varphi_e^+ = \varphi_{d_3} = \varphi_{u_1} = \varphi_{u_2} = \varphi_0$$

with φ_0 arbitrary. In the scattering event we are discussing, φ_0 equals 0 "before" and $\sqrt{\pi}/2$ "after". The displaced vacuum can be thought of as a collective state of the fermi vacuum which is accessible because the fermions are massless.

If we turn the fermion masses back on, the continuous vacuum degeneracy is lifted and we find, in the simplest case, that the allowed values of φ_0 are integer multiples of $\sqrt{\pi}$ (corresponding to the successive minima of the $\cos 2\sqrt{\pi}\varphi$ mass terms). Therefore the displaced vacuum corresponding to $\varphi_0=\sqrt{\pi}/2$ is not stable: it lies precisely midway between the stable vacua $\varphi_0=0$ and $\varphi_0=\sqrt{\pi}$ and must eventually decay to one or the other of them.

This vacuum decay process allows the final state to sort itself out into full solitons which can be interpreted as ordinary fermions. Representative configurations corresponding to the two alternatives $\varphi_0=\sqrt{\pi}/2\rightarrow\varphi_0=\sqrt{\pi}$ and $\varphi_0=\sqrt{\pi}/2\rightarrow\varphi_0=0$ are shown in Figs.3 and 4.

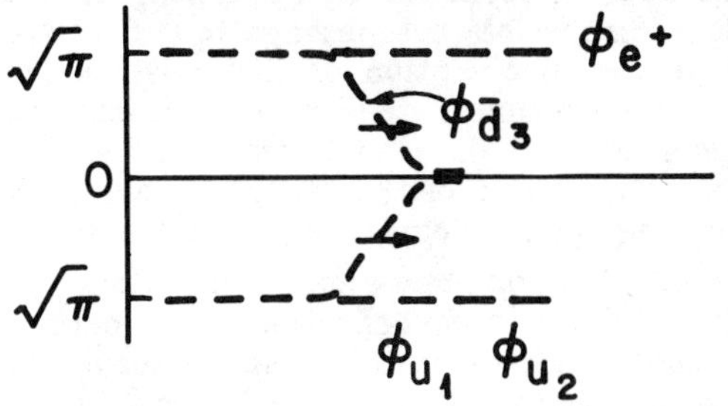

Figure 3. A typical $\Delta B=1$ outgoing state.

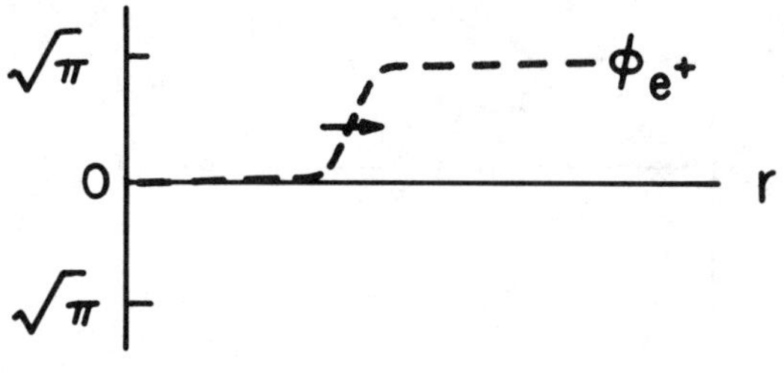

Figure 4. A typical $\Delta B=0$ outgoing state.

In fermion language, the two alternatives correspond to

$$e^+{}_L + M \rightarrow M + u_{1_R}\, u_{2_R}\, d_{3_L}$$

and

$$e^+{}_L + M \rightarrow M + e^+{}_R$$

respectively. The first option corresponds to $\Delta B=1$ while the second corresponds to $\Delta B=0$.

Within each of the two classes there are a large number of accessible final states and we have explicitly identified only the simplest state in each class. We expect not to be able to use classical arguments to calculate the partial cross-section into into each particular channel: the transition to states with definite fermion content is caused by the mass terms, and the mass terms are precisely the non-soluble interaction terms of the model. On the other hand, since there is no obvious distinction between the time histories corresponding to decay to $\varphi_0=0$ and $\varphi_0=\sqrt{\pi}$, respectively, it seems reasonable to expect that the total cross-sections for $\Delta B=0$ and $\Delta B=1$ final states should be roughly equal. This fits very nicely with our previous argument and again indicates that the calculation of the detailed S-matrix is a non-trivial quantum field theory problem.

At this point there is an interesting side remark to be made concerning the role of conservation laws. Consider a typical $\Delta B=1$ process:

$$e^+{}_L + M \to M + u_{1_R}\, u_{2_R}\, d_{3_L}$$

The $SU_3^c x U_1^{em}$ charges of the fermions are conserved as always: they could only have disappeared onto the monopole core and there is a Coulomb energy barrier to forbid that. However, weak isospin is not conserved in this process ($I_3^w=0 \to I_3^w=\frac{1}{2}$) despite our demonstration that it is conserved in the underlying short-distance scattering from the monopole. In fact there is no accessible final state with a definite number of fermions which conserves weak isospin! But, according to our previous discussion, the passage to a final state with definite fermion content requires the intervention of the mass terms and the mass terms explicitly violate weak isospin. Instead of disappearing at very short distances onto the monopole core, weak isospin disappears at the large distance and low energy scales characteristic of the fermion masses. For that reason, SU_2^w gauge fields are not required to come into play in the scattering process.

As we remarked earlier, the "one-and-a-half generation" model specified by the boundary conditions of (4.4) is perhaps the most realistic model of low-energy scattering from the monopole. In this case, the scattering of a positron from the monopole (in the approximation of neglecting fermion masses) produces an out state containing "third-integral" solitons as shown in Fig.5.

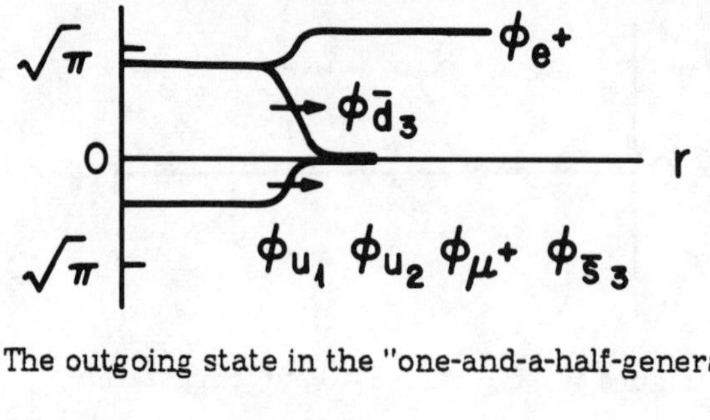

Figure 5. The outgoing state in the "one-and-a-half-generation" model.

The line of argument used to interpret the previous "half-soliton" results now indicates that there are three equally probable classes of final states having the net quantum numbers of e^+, $d_3 u_1 u_2$ and $d_3 \bar{s}_3 \mu^+$, respectively. The addition of one more active doublet reduces the total baryon-number-violating cross-section slightly, compared to the one-generation case, but allows for the occurence of a new kind of lepton-number-violating channel. The generalization of these arguments to higher energies where more fermion doublets are effectively massless is straightforward.

CONCLUSIONS

Clearly, the problem of calculating the monopole catalysis S-matrix is far from being solved. The preceeding discussion was meant only to argue that, on the basis of very simple considerations, it is possible to make plausible guesses about some general properties of the S-matrix. The same discussion makes it clear that the calculation of the details of the S-matrix is a non-trivial quantum field theory problem. It may be that numerical techniques, a la Monte Carlo simulations, will be needed to make progress here. This will certainly be true if one wants information about scattering to specific hadron final states!

There are also some general issues which are imperfectly understood and which ought to be looked into more carefully. We have emphasized the role of conservation laws in the scattering process, but have restricted our discussion to the behavior of "diagonal" generators of symmetry groups. This is not because the off-diagonal, non-Abelian generators are unimportant, but because our bosonization techniques are not particularly well-suited to their discussion. We have also seen that mass terms play an important role in the understanding of the "fractional soliton"" outcome of typical scattering events in the bosonized picture. For simplicity, we have assumed that the fermion mass is a scale-independent quantity. It has, however, been pointed out [5] that the monopole disturbs the hierarchy structure which protects some of the five-dimensional Higgs fields from having super-large expectation values. This in turn means that the fermion masses will in general become large in the immediate neighborhood of the monopole core. Since there is a small coupling constant (the Yukawa coupling of the Higgs and fermi fields) involved, there is no reason to believe that our general picture is affected by this phenomenon, but the question has yet to be investigated in detail. This is a particular aspect of the general question whether the simplified models used to discuss the phenomenon of monopole catalysis of baryon decay are fully representative of the complexity of realistic grand unification theories. The last word on this as well the other subjects discussed here is probably not yet in!

REFERENCES

1. V.A. Rubakov, Nucl. Phys. B203, 311 (1982); ZhETF Pis'ma 33, 658 (1981).

2. C.G. Callan, Jr., Phys. Rev. D25, 2141 (1982); Phys. Rev. D26, 2058 (1982); Nucl. Phys. B204 (1982).

3. This issue was first raised with me by F. Wilczek. The same question is raised in a Rockefeller University preprint by Grossman, Lazarides and Sanda.

4. C.G. Callan, Jr., "Monopole Catalysis of Baryon Decay," Nov. 1982 Princeton preprint, to be published in the Proceedings of the Wingspread Conference on Magnetic Monopoles, Carrigan and Trower eds.

5. This peculiar half-soliton behavior was first pointed out to me by E. Witten.

6. Private communication from E. Witten and J. Harvey.

DIMENSIONAL TRANSMUTATION IN BROKEN SUPERGRAVITY*

J. Ellis and J.S. Hagelin
Stanford Linear Accelerator Center
Stanford University, Stanford, California 94305

D. V. Nanopoulos
Theory Division, CERN, Geneva, Switzerland

K. Tamvakis
Physics Department, University of Ioannina, Ioannina, Greece

Presented by John Ellis

ABSTRACT

Weak interaction gauge symmetry breaking can be generated by radiative corrections in a spontaneously broken supergravity theory, provided the top quark is heavy enough. In one class of such theories the weak Higgs vacuum expectation values are determined by dimensional transmutation à la Coleman-Weinberg, and may be considerably larger than the magnitudes of susy breaking mass parameters. In this scenario $m_t \geq 65\ GeV$, the supersymmetric partners of known particles may have masses $\ll m_W$, the mass of the lighter neutral scalar Higgs boson is determined by radiative corrections, and there is some variant of a light pseudoscalar axion. In contrast to conventional Coleman-Weinberg models, the weak phase transition is second order and there is no likelihood of excess entropy production.

1. INTRODUCTION

Supersymmetry (susy) has recently attracted considerable phenomenological attention because[1] it can protect the weak interaction scale and preserve the hierarchy $m_W/m_P \ll 1$. However, susy does not by itself predict or explain the magnitude of m_W. Also, although the susy partners of many familiar particles must have masses $\leq O(1)\ TeV$ if the hierarchy is to be maintained, the primordial susy breaking scale $\sqrt{d}$ could be much larger.[2,3] Global susy may be broken either in the gauge sector (D breaking) or in the chiral sector (F breaking). Neither of these schemes has proved completely satisfactory: no completely realistic D breaking model exists, and F breaking models seem rather contrived. Scenarios have been proposed in which the weak interaction scale is obtained from high order radiative corrections,[3] with symmetry breaking driven by a heavy top quark.[4,5] When $\sqrt{d} \geq O(10^{11})\ GeV$ it seems essential to consider the effects of local susy, since the gravitino mass $m_{3/2} = O(d/m_P) = O(m_W)$,

*Work supported by the Department of Energy, contract DE-AC03-76SF00515.

and scalar fields acquire contributions m to their masses of $O(m_{3/2})$.[6] There is now some consensus among susy model-builders that the super-Higgs mechanism in $N = 1$ supergravity theories may now be the work promising mechanism for susy breaking. Some phenomenological supergravity models have been proposed in which weak gauge symmetry breaking is realized at the tree level.[7] However, it seemed[8] to us more natural to suppose that radiative corrections play an important role, possibly with a heavy t quark driving weak gauge symmetry breaking, as had been proposed earlier[3,4,5] in the context of global susy (see also Ref. 9). Moreover, there emerged[10] difficulties with alternative models for weak symmetry breaking which employed light singlet chiral superfields. In the previous paper[8] we demonstrated the feasibility of a similar scenario in the context of local susy, without solving the full coupled set of renormalization group equations for the susy breaking parameters.

2. SYMMETRY BREAKING BY RADIATIVE CORRECTIONS

Conveniently enough, the full renormalization group equations for these parameters are available from a previous analysis[5] in the context of global susy. All that is necessary in order to arrive at an analogous broken supergravity model is to choose a somewhat different set of initial conditions for the susy breaking parameters.[11] These include gaugino masses M,[8] scalar boson masses m,[6] and trilinear scalar couplings λ.[7,12,13] One's guess might be that all of these parameters are $O(m_{3/2})$. However, it has been proposed[14] on the basis of a $U(n)$ symmetry among the chiral superfields respected by perturbative gravitational effects, that perhaps $M = O(\alpha/2\pi)m_{3/2}$.[13] We see no particular reason why such a symmetry should survive non-perturbative gravitational effects, and it is in any case broken by Yukawa couplings which may be large for the top quark. Therefore we prefer to retain $\hat{M} \equiv M/m = O(1)$. The initial value of the ratio $\hat{\lambda} \equiv \lambda/m$ is related[7,12,13] to unknown parameters of a hidden sector of the theory, and is model-dependent but probably $O(1)$.

We prefer to keep an open mind about this sector of the theory, which may well not be a simple polynomial in a single unknown chiral superfield added on to the superpotential for known chiral superfields,[15] but may reflect some more complicated dynamics at scales $O(m_P)$. In addition to the mass parameters listed above, the low energy Higgs potential involving two Higgs superfields $H_{1,2}$ with susy breaking masses $m_{1,2}$ may also include a quadratic term H_1H_2 with coefficient $\mu \times O(m_{3/2})$ related to a quadratic term $\propto H_1H_2$ in the chiral superpotential. There is no *a priori* connection between the values of μ and of $m_{3/2}$, and if $\mu \ll m_W$ the physical Higgs spectrum contains an axion. Phenomenological model-builders search in the multi-dimensional space of the parameters m, $\hat{M}$, $\hat{\lambda}$, μ and the t quark Yukawa coupling h_t to the Higgs H_2 for outputs of the renormalization group equations in which m_2^2 has been driven negative by h_t, permitting the breakdown of $SU(2) \times U(1)$ to $U(1)_{em}$. Typically, for given choices of m, $\hat{M}$, $\hat{\lambda}$ and μ we find a range of values of h_t which give m_2^2 negative, corresponding to $m_t \geq O(M_W)$. Since m_2^2 varies quite rapidly as one approaches the strong interaction scale, different negative values of m_2^2 are attained at the price of modest variations in h_t and hence m_t.

3. DIMENSIONAL TRANSMUTATION

The results of one general analysis[16] of this parameter space are reported at this meeting by Ibáñez. Here we discuss some plausible hypotheses which diminish the dimensionality of the parameter space and constrain the theory in an interesting way. Since μ has no definite reason to be $O(m_W)$, and could well be much less, perhaps $O((\alpha/\pi)^n)m_W$ or $O(m_W^2/m_X)$ or even zero, we consider the possibility

$$\mu = 0, \quad \textit{or at least} \quad \ll m_W \,. \tag{1}$$

In this case the weak gauge symmetry breaking occurs near a scale Q_0 where the linear combination $m_1^2 + m_2^2$ of Higgs mass^2 parameters vanishes. This scale Q_0 is independent of m as long as $m \ll Q_0$. Furthermore, for a given choice of $\hat{M}$ and $\hat{\lambda}$ there is a *unique* value of h_t and hence m_t which fixes Q_0 so as to give m_W correctly. This enables us to *predict* m_t as a function of $\hat{M}$ and $\hat{\lambda}$, and we find that for all plausible values of these parameters

$$m_t \geq 65 \; GeV \,. \tag{2}$$

In contrast to other models, in this scenario the unseen supersymmetric partners of known particles could be lurking arbitrarily close to the present experimental lower limits on their masses. In this scenario the weak interaction scale is divorced from the scalar and gravitino masses, since it is fixed by dimensional transmutation in the style of Coleman and E. Weinberg[17] The difference is that whereas in their case it was the logarithmic evolution of a *quartic* Higgs coupling which determined the weak interaction scale, in our susy case it is the logarithmic evolution of a *quadratic* Higgs coupling. As in the Coleman-Weinberg analysis, we have a light neutral Higgs scalar whose mass is determined by radiative corrections, and we also have the pseudoscalar axion mentioned earlier. We assume that this axion could ultimately be made phenomenologically acceptable, perhaps by becoming a new improved invisible axion in a more complicated model[18,19] or perhaps by μ being sufficiently large ($\geq O(1)MeV$) to push the axion mass $m_a = O(\mu m)^{1/2}$ above the experimental lower limit of 350 MeV from $K \rightarrow \pi + a$ decay. There may also be a constraint on μ from cosmology which depends on the relative masses of susy particles, and on that of the photino in particular. It is interesting to speculate that the initial stage of GUT symmetry breaking could also be driven by radiative corrections, in which case one might hope to understand why $m_W/m_X \ll m_X/m_P \ll 1$ along the lines proposed in Ref. 20. In this connection we make some remarks about the variation in couplings and mass parameters between m_P and m_X.

We assume there are no other light chiral superfields besides the Higgses $H_{1,2}$, the quarks and the leptons. Therefore the low energy potential for the neutral Higgses is[21]

$$V = \frac{g_2^2 + g'^2}{8}(|H_1|^2 - |H_2|^2)^2 + m_1^2|H_1|^2 + m_2^2|H_2|^2 - m_3^2(H_1H_2 + H_1^*H_2^*) \quad (3)$$

The quartic D-term allows the Higgses to leak to infinity unless[21]

$$m_1^2 + m_2^2 > 2m_3^2 \quad (4)$$

and there is $SU(2) \times U(1)$ breaking if[21]

$$m_3^4 > m_1^2 m_2^2 \quad (5)$$

with

$$\frac{v_1}{v_2} \equiv \frac{< 0|H_1|0 >}{< 0|H_2|0 >} = \cot\theta : \sin 2\theta = \frac{2m_3^2}{(m_1^2 + m_2^2)} \quad . \quad (6)$$

We assume that H_2 gives mass to the t quark $m_t = (1/\sqrt{2})h_t v_2$, and $h_t > h_b$ so that the renormalization group drives $m_2^2 < m_1^2$ at present energies, and we will be interested in what happens when $m_3^2 = O(\mu m) \to 0$. In leading order of the renormalization group equations the Higgs mass parameters m_i^2 in the effective potential depend (logarithmically) only on the corresponding $|H_i|^2$, and they are positive at large scales ensuring that condition (4) is obeyed. If $m_1^2 + m_2^2$ decreases to zero at some scale $|H_i| = Q_0$, this will determine the value of $v^2 \equiv v_1^2 + v_2^2$, while

$$\delta^2 \equiv v_2^2 - v_1^2 = \frac{2(m_1^2 - m_2^2)}{(g_2^2 + g'^2)} = \frac{4m_1^2(Q_0^2)}{(g_2^2 + g'^2)} \quad . \quad (7)$$

The combination $m_1^2 + m_2^2$ becomes negative at scales less than Q_0, resulting in the form of potential shown in Fig. 2. If $m_{1,2}$ are much less than the dimensional transmutation scale Q_0 then equation (7) tells us that the absolute minimum of the potential is at

$$v_1^2 \approx v_2^2 \approx \frac{v^2}{2} \quad (8)$$

and

$$Q_0 = \sqrt{\frac{e}{2}}\, v \approx 290\ GeV\ . \quad (9)$$

To calculate the scale Q_0 at which $m_1^2 + m_2^2 = 0$ we need the leading order renormalization group equations of Ref. 5 which are valid for $Q \gg M^2, m^2$. We have in their notation the initial conditions

$$m_3 = m_4 = m_5 = m_7 = m_9 = 0; \ m_6 = m_8 = m_{10} = \hat{\lambda}\, m \quad (10)$$

In the limit that $m = 0$ our initial conditions become a limiting case of those considered in Ref. 5, with the only susy breaking in the initial conditions coming from $M \neq 0$. Neglecting all Yukawa couplings except those of the top quark, the relevant renormalization group equations are

$$Q\frac{dm_1^2}{dQ} = \frac{1}{(4\pi)^2}[-6g_2^2M_2^2 - 2g'^2M_1^2] \tag{11a}$$

$$Q\frac{dm_2^2}{dQ} = \frac{1}{(4\pi)^2}[-6g_2^2M_2^2 - 2g'^2M_1^2 + 6h_t^2(m_{q_3}^2 + m_{p_3}^2 + m_2^2 + m_{10}^2)] \tag{11b}$$

$$Q\frac{dm_{10}}{dQ} = \frac{1}{(4\pi)^2}\left[\frac{-32}{3}g_3^2M_3 - 6g_2^2M_2 - \frac{26}{9}g'^2M_1 + 6h_t^2m_{10}\right] \tag{11c}$$

$$Q\frac{dm_{q_3}^2}{dQ} = \frac{1}{(4\pi)^2}\left[\frac{-32}{3}g_3^2M_3^2 - 6g_2^2M_1^2 - \frac{2}{9}g'^2M_1^2 \right.$$

$$\left. + 2h_t^2(m_{q_3}^2 + m_{p_3}^2 + m_2^2 + m_{10}^2)\right] \tag{11d}$$

$$Q\frac{dm_{p_3}^2}{dQ} = \frac{1}{(4\pi)^2}\left[\frac{-32}{3}g_3^2M_3^2 - \frac{32}{9}g'^2M_1^2 + 4h_t^2(m_{q_3}^2 + m_{p_3}^2 + m_2^2 + m_{10}^2)\right] \tag{11e}$$

$$Q\frac{dm_{n_3}^2}{dQ} = \frac{1}{(4\pi)^2}\left[\frac{-32}{3}g_3^2M_3^2 - \frac{8}{9}g'^2M_1^2\right] \tag{11f}$$

for the susy breaking scalar mass parameters, and

$$Q\frac{dh_t}{dQ} = \frac{h_t}{(4\pi)^2}\left[\frac{-16}{3}g_3^2 - 3g_2^2 - \frac{13}{9}g'^2 + 6h_t^2\right] \tag{12}$$

for the t quark Yukawa coupling. The gaugino masses are

$$M_{3,2} = \frac{g_{3,2}^2(Q^2)M}{g_{GUT}^2}, \quad M_1 = \frac{5}{3}\frac{g'^2(Q^2)M}{g_{GUT}^2} \tag{13}$$

while $g_{3,2}$ and g' evolve conventionally with Q.

We have integrated these renormalization group equations for different starting values of the ratios $\hat{M}$ and $\hat{\lambda}$, and located the corresponding values of m_t which yield a dimensional transmutation scale $Q_0 = 290\ GeV$. Vacuum stability conditions prefer[22] $\hat{\lambda} < 3$, but this condition should be interpreted *cum grano salis*. It is applicable at scales $O(m_W)$ where $\hat{\lambda}$ is renormalized from its initial value in different ways for different trilinear couplings. Finite temperature effects in the early universe favour the conventional local minimum. Tunnelling into other minima is suppressed by $\exp(-O(1)/h^2)$ where h is the relevant Yukawa coupling. The false vacuum is more stable than the age of the universe except perhaps for transition to the minimum controlled by h_t. If $m_2^2 \ll m_{q_3}^2, m_{p_3}^2$ at scales $O(m_W)$, the absolute stability condition[22] on $\hat{\lambda}$ is modified to $\hat{\lambda}_t \equiv \lambda_t(m_W)/m_{\tilde{t}}(m_W) < 2$. This condition is obeyed if the initial $\hat{\lambda} \leq 2(1/2)$, as can be seen in the Table. Even if this condition is not obeyed, it is still possible that the lifetime of the false vacuum may be longer than the age of the universe for relevant values of h_t.

TABLE

Masses in models with the SU(2) × U(1) breaking scale determined by radiative corrections.

	$\hat{M} = \hat{\lambda} = 1$	$M = 1$, $m = \lambda = 0$	$\hat{M} = .35$, $\hat{\lambda} = 1$	$\hat{M} = 1$, $\hat{\lambda} = 2.5$	$\hat{\lambda} = \hat{m}_T = \sqrt{3/2}$, $M = m_{\overline{F}}$
m_t	88	67	140	70	82
$\hat{m}_{H_o}$	.67	.46	.26	.66	.67
$\hat{m}_{\tilde{q}_3}$	2.7	2.6	1.1	2.7	2.7
$\hat{m}_{\tilde{q}_{1,2}}$	2.9	2.7	1.4	2.9	3.0
$\hat{m}_{\tilde{p}_3}$	2.3	2.4	.60	2.3	2.4
$\hat{m}_{\tilde{p}_{1,2}}$	2.8	2.6	1.4	2.8	2.9
$\hat{m}_{\tilde{n}_{1,2,3}}$	2.8	2.6	1.4	2.8	2.8
$\hat{m}_{\tilde{\ell}_{1,2,3}}$	1.2	.73	1.0	1.2	1.2
$\hat{m}_{\tilde{e}_{1,2,3}}$	1.1	.39	1.0	1.1	1.3
$\hat{\lambda}_t$	1.6	1.5	1.0	2.2	1.7

All masses denoted $\hat{m}_i$ are in units of the $\underline{5}$-plet scalar masses at the grand unification scale m_X, except that masses in the second column are in units of the gaugino mass at the scale m_X.

4. RESULTS

Our results for m_t are shown in Fig. 1: they were determined by integrating the renormalization group equation for h_t down to a momentum scale $Q = m_t$. Note that we are not able to find solutions if $\hat{M} < 0.35$ for $\hat{\lambda} = 1$. Within the

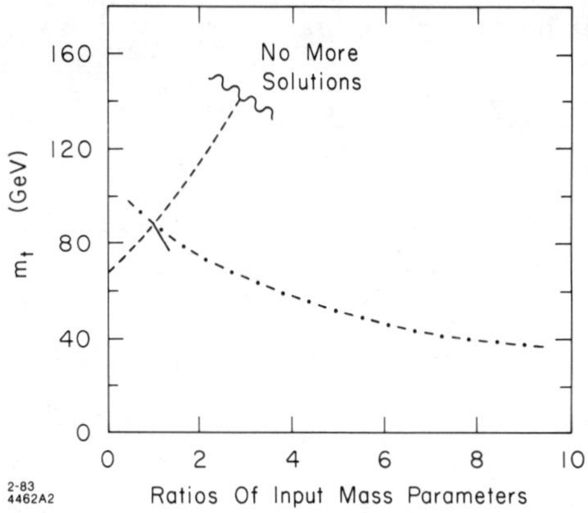

Fig. 1. Predictions of m_t corresponding to different values of the input mass ratios $\hat{M}^{-1} \equiv m/M$ (dashed line), $\hat{\lambda} \equiv m_{10}/m$ (dashed-dotted line) and $m_T/m_{\bar{F}}$ (solid line).

allowed range of $\hat{M}$ we find $m_t \geq 65$ *GeV* in the supersymmetric Coleman-Weinberg scenario for $\hat{\lambda} < 2\frac{1}{2}$. If m_t turns out to be < 65 *GeV*, our scenario could still apply if our present vacuum is unstable, or if there is a fourth generation. In general, m_2^2 is evolving very rapidly at low Q, which means that the values of m_t needed are not much larger than the typical ranges found when we look for general solutions [5,16] to the inequalities (4,5) rather than looking specifically for $m_1^2 + m_2^2 \to 0$. In the general case we often find $v_1 \ll v_2 \approx v$, so that the same value of h_t gives m_t a factor $\sqrt{2}$ larger than in the dimensional transmutation case (8).

The rapid final stages of evolution of m_2^2 are driven by the increases in the t quark Yukawa coupling and more importantly in the squark masses which occur when $g_3^2/4\pi$ becomes large. Thus in the supersymmetric Coleman-Weinberg scenario the weak interaction scale is related to that of the strong interactions, while the absolute values of m and $m_{3/2}$ are not directly related to m_W. This contrasts with what usually happens in models of weak gauge symmetry breaking in supergravity models[7,8,9,16] where m_W is connected with m and $m_{3/2}$, but is not directly related to the strong interaction scale. In practice, phenomenology dictates that m must be large enough for all unobserved particles to have been able to escape detection, but it could be as low as 15 *GeV* in our scenario, thus offering the prospect of imminent detection of susy particles. The table shows values of the physical masses of these particles in units of m for selected representative values of the input parameters $\hat{M}$ and $\hat{\lambda}$. We see that the lightest spin-zero superpartners are the sleptons. For small M the lightest gaugino is approximately a photino $\tilde{\gamma}$ with mass

$$m_{\tilde{\gamma}} \approx \frac{g'^2 M_2 + g_2^2 M_1}{(g_2^2 + g'^2)} \approx \frac{8}{3} \frac{g_2^2 g'^2}{(g_2^2 + g'^2) g_{GUT}^2} M \approx 0.47 M \,. \tag{14}$$

This could be light enough to be pair-produced at PEP and PETRA, and the selectron mass could well be small enough for the cross-section for $e^+e^- \to \tilde{\gamma}\tilde{\gamma}\gamma$ to be detectably large at present energies.[23] Turning now to the physical Higgs bosons in this class of model,[21] the charged bosons $H^\pm$ and the heavier neutral scalar boson $H^{0\prime}$ acquire masses

$$m_{H^\pm} = m_{W^\pm}, \qquad m_{H^{0\prime}} = m_{Z^0} \tag{15}$$

at the tree level. The lighter extra scalar boson H^0 acquires

$$m_{H^0}^2 = \frac{1}{16\pi^2}\left[6h_t^2(m_{q_3}^2 + m_{p_3}^2 + m_2^2 + m_{10}^2) - 12g^2M_2^2 - 4g'^2M_1^2\right] \qquad (16)$$

from radiative corrections. Values of m_{H^0} corresponding to typical values of the input parameters $\hat{M}$ and $\hat{\lambda}$ are also given in the table. Typically

$$m_{H^0} \approx \left(\frac{1}{4} \text{ to } 2/3\right) m \qquad (17)$$

which is not much smaller than the slepton masses, as a result of the relatively large squark masses exhibited in the table and appearing in eq. (16). Finally, our spectrum contains a light neutral pseudoscalar axion state which must be exorcised in one of the ways discussed earlier. This can[19] be done in such a way as to avoid astrophysical and cosmological pitfalls. Our class of susy Coleman-Weinberg models also avoids the danger[24] of excess entropy generation during the weak phase transition, because as seen from Fig. 2 the origin is an unstable extremum and there is a second order phase transition once the temperature falls below $O(m)$.

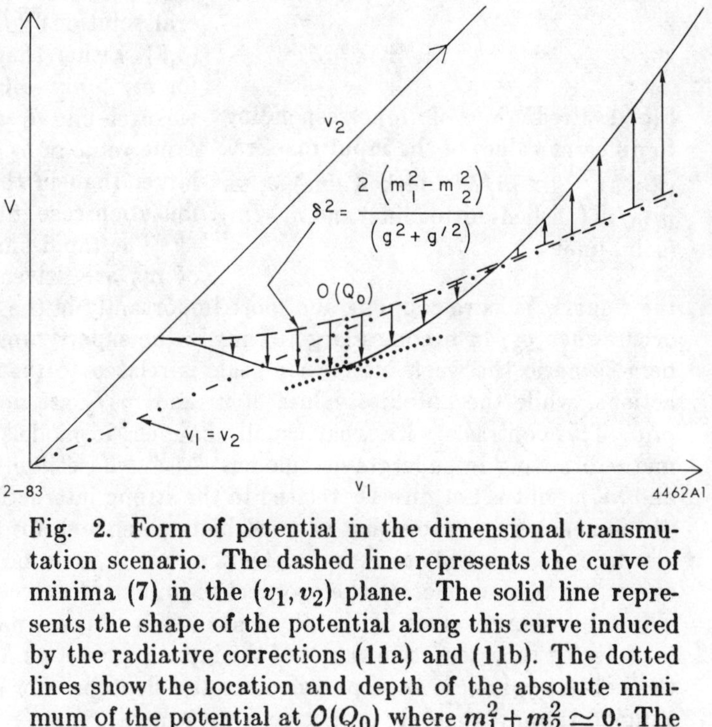

Fig. 2. Form of potential in the dimensional transmutation scenario. The dashed line represents the curve of minima (7) in the (v_1, v_2) plane. The solid line represents the shape of the potential along this curve induced by the radiative corrections (11a) and (11b). The dotted lines show the location and depth of the absolute minimum of the potential at $O(Q_0)$ where $m_1^2 + m_2^2 \simeq 0$. The extremum at $v_1 = 0, v_2 \neq 0$ is unstable since $m_1^2 + m_2^2 < 0$ at scales $O(m) \ll Q_0$.

5. A SUPERSYMMETRIC HIERARCHY OF HIERARCHIES?

Before closing we would like to add a few comments about the possibility of embedding this susy Coleman-Weinberg scenario in a GUT. One remark contains the initial values of the scalar masses that we have assumed. There is no good reason why the masses of $\underline{\bar{5}}$ and $\underline{10}$ matter fields F and T should be the same at the GUT breaking scale m_X, nor why the $\underline{5}$ and $\underline{\bar{5}}$ Higgs masses should be the same. Even if some symmetry fixed them to be equal at m_P, they would differ at M_X. We have evaluated this possible difference in the minimal $SU(5)$ GUT[1] and found that

$$1 \leq \hat{m}_T^2 \equiv \frac{m_T^2(m_X)}{m_{\bar{F}}^2(m_X)} \leq 1.5 \tag{18}$$

with $m_{H_1}^2 \approx m_{H_2}^2 \approx m_{\bar{F}}^2$. Figure 1 shows that variation in the range (18) does not have a substantial effect on the required t quark mass, though it can increase the physical masses of squarks and sleptons from the $\underline{10}$ representations of $SU(5)$, such as the $\tilde{e}_R$, $\tilde{\mu}_R$ and $\tilde{\tau}_R$.

It is enticing to speculate whether the grand unification scale m_X could also be determined by dimensional transmutation, thanks to some susy breaking scalar mass in the GUT sector being driven to zero at a scale $Q = O(m_X)$. This would be a reincarnation of the double Coleman-Weinberg scenario of Ref. 20, in which the "hierarchy of hierarchies" $m_W/m_X \ll m_X/m_P \ll 1$ was ascribed to the rapid evolution of the couplings of large GUT representations such as the $\underline{24}$ of Higgs in $SU(5)$ which gave a very large dimensional transmutation scale to the GUT breaking. This suggestion would now be applied to the susy breaking mass parameters instead of the quartic scalar couplings as illustrated in Fig. 3. Unfortunately, such a scenario cannot be realized in the minimal susy GUT[25] where the lightness of the Weinberg-Salam Higgses and the heaviness of their colour triplet partners are enforced by the fine-tuning of two mass parameters in the superpotential. If one supplements the conventional minimal $SU(5)$ GUT with additional $\underline{40}$ and $\overline{\underline{40}}$ chiral superfields with a coupling ν to the adjoint $\underline{24}$ of Higgs, one can easily find plausible initial conditions at m_P which can drive m_{24}^2 to zero at scales $Q = O(10^{-3})m_P$, such as

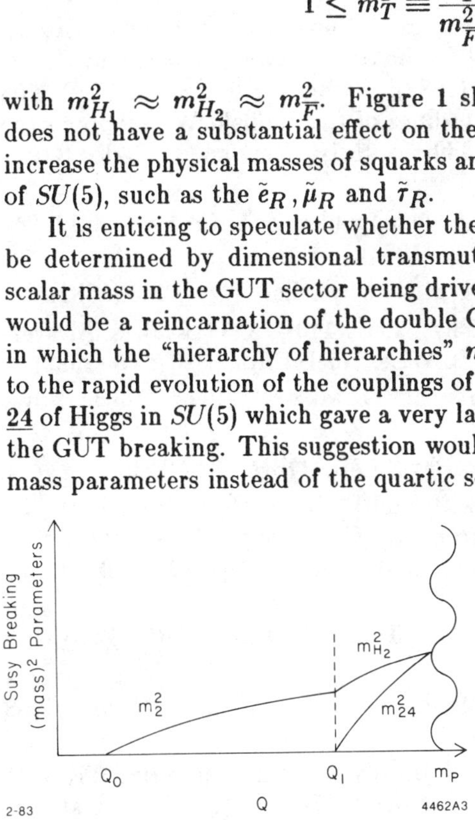

Fig. 3. Qualitative features of the variation of susy breaking mass parameters in the "hierarchy of hierarchies" scenario. It may be possible to generate $m_X = O(Q_1)$: $Q_0/m_P \ll Q_1/m_P \ll 1$.

$$\frac{g_5^2}{4\pi} = 0.19 \; ; \; \frac{\nu^2}{4\pi} = 0.004 \; ; \; m_{40}^2 = m_{\overline{40}}^2 = other\ m^2 \; ; \; M = O(2)\, m \; . \tag{19}$$

It remains to find a cleverer model featuring such a supersymmetric hierarchy of hierarchies in which the Higgs doublet/triplet splitting problem is also solved.

ACKNOWLEDGEMENTS

We would like to thank R. A. Flores, L. Hall, L. E. Ibáñez, J. Polchinski, M. A. Sher and M. B. Wise for useful discussions, and express our gratitude to G. Farrar and F. Henyey for their kind hospitality at this meeting.

REFERENCES

1. For reviews, see: D. V. Nanopoulos – "Supersymmetry versus Experiment Workshop", ed. D. V. Nanopoulos, A. Savoy-Navarro and C. Tao – CERN TH-3311/EP-82/63 (1982), p. 99; J. Ellis, SLAC-PUB-3006 (1982), to appear in the Proceedings of the Nuffield Workshop on the Very Early Universe, ed. G. Gibbons, S. Hawking and S. Siklos (Cambridge University Press, 1983).
2. R. Barbieri, S. Ferrara and D. V. Nanopoulos, Zeit. für Phys. *C*13, 267 (1982) and Phys. Lett. 116B, 16 (1982).
3. J. Ellis, L. E. Ibáñez and G. G. Ross, Phys. Lett.113*B*, 283 (1982).
4. L. E. Ibáñez and G. G. Ross, Phys. Lett. 110*B*, 215 (1982); L. Alvarez-Gaumé, M. Claudson and M. B. Wise, Nucl. Phys. *B*207, 96 (1982).
5. K. Inoue, A. Kakuto, H. Komatsu and S. Takeshita, Prog. Theor. Phys. 68, 927 (1982).
6. J. Ellis and D. V. Nanopoulos, Phys. Lett. 116*B*, 133 (1982).
7. A. H. Chamseddine, R. Arnowitt, and P. Nath, Phys. Rev. Lett. 49, 970 (1982); R. Barbieri, S. Ferrara and C. A. Savoy, Phys. Lett. 119*B*, 343 (1982); L. E. Ibáñez, Phys. Lett. 118*B*, 73 (1982); talk at this meeting by R. Arnowitt.
8. J. Ellis, D. V. Nanopoulos and K. Tamvakis, Phys. Lett. 121B, 123 (1983).
9. L. E. Ibáñez, Universidad Autónoma de Madrid preprint FTUAM/82-8 (1982).
10. H. P. Nilles, M. Srednicki and D. Wyler, CERN preprint TH-3461 (1982); A. B. Lahanas, CERN preprint TH-3467 (1982); talk at this meeting by H. P. Nilles.
11. E. Cremmer, B. Julia, J. Scherk, S. Ferrara, L. Girardello and P. van Nieuwenhuizen, Nucl. Phys. *B*147, 105 (1979); E. Cremmer, S. Ferrara, L. Giradello and A. van Proeyen, Phys. Lett. 116B, 231 (1982). and CERN preprint TH-3348 (1982).
12. H. P. Nilles, M. Srednicki and D. Wyler, Phys. Lett. 120*B*, 346 (1982).
13. L. Hall, J. Lykken and S. Weinberg, University of Texas preprint UT-TG-1-83 (1983); talk at this meeting by S. Weinberg.
14. M. K. Gaillard, LBL preprint 14647 (1982); R. Arnowitt, A. H. Chamseddine and P. Nath, Harvard preprint HUTP-82/A055-MIB2583 (1982); S. Weinberg, University of Texas preprint UTTG-2-82 (1982).

15. J. Polonyi, Budapest preprint KFKI-1977-93 (1977).

16. L. Alvarez-Gaumé, J. Polchinski and M. B. Wise, Harvard preprint HUTP-82/A063 (1982); L. E. Ibáñez and C. López, Universidad Autónoma de Madrid preprint FTUAM/83-2 (1983); talk at this meeting by L. E. Ibáñez.

17. S. Coleman and E. Weinberg, Phys. Rev. *D7*, 1888 (1973).

18. M. Dine. W. Fischler and M. Srednicki, Phys. Lett. 104*B*, 199 (1981).

19. P. Sikivie, Phys. Rev. Lett. 48, 1156 (1982), J. Preskill, M. B. Wise and F. Wilczek, Phys. Lett. 120*B*, 127 (1982); L. Abbott and P. Sikivie, Phys. Lett. 120*B*, 133 (1982); M. Dine and W. Fischler, Phys. Lett. 120*B*, 137 (1982).

20. J. Ellis, M. K. Gaillard, A. Peterman and C. T. Sachrajda, Nucl. Phys. *B*164, 253 (1980).

21. K. Inoue, A. Kakuto, H. Komatsu and S. Takeshita, Prog. Theor. Phys. 67, 1889 (1982); see also R. A. Flores and M. Sher, UCSC preprint TH-154-82 (1982).

22. J.-M. Frère, D.R.T. Jones and S. Raby, University of Michigan preprint UMHE 82-58 (1982).

23. P. Fayet, Phys. Lett. 117*B*, 460 (1982); J. Ellis and J. S. Hagelin, SLAC-PUB-3014 (1982).

24. E. Witten, Nucl. Phys. *B*177, 477 (1981); see also R. A. Flores and M. Sher, Ref. 21.

25. S. Dimopoulos and H. Georgi, Nucl. Phys. *B*193, 150 (1981); N. Sakai, Zeit. für Phys. *C*11, 153 (1982).

NON-LINEAR REPRESENTATIONS OF EXTENDED SUPERSYMMETRY, HIGGS AND SUPERHIGGS EFFECT

S. Ferrara
CERN, Geneva, Switzerland

ABSTRACT

We discuss non-linear realizations of extended supersymmetries with central charges. Their coupling to N-extended supergravity provides a model-independent discussion of the Higgs and superHiggs effect in higher N locally supersymmetric theories.

Non-linear Lagrangians describing chiral dynamics have been widely considered in the past[1)] to describe low-energy spontaneously broken chiral symmetries and in particular the low-energy properties of pion physics. When applied to N = 1 supersymmetry, non-linear representations of supersymmetry summarize the low-energy properties of goldstino amplitudes and contain all the information of supercurrent algebra, i.e., the way in which the goldstino, the spin-1/2 goldstone fermion of spontanesouly broken supersymmetry, couples to ordinary matter. The theoretical set-up for the construction of non-linear realizations of supersymmetry and their consequent invariant Lagrangian was given several years ago by Volkov and Akulov[2)] and further developed in Ref. 3).

When spontaneously broken Lagrangians are coupled to supergravity, the goldstino is eaten up by the spin 3/2 gravitino, the gauge fermion of local supersymmetry, and one obtains model independent relations between the gravitino mass and the scale vacuum energy of a global system which results from the spontaneous breaking of supersymmetry. This relation, as shown by Deser and Zumino[4)], is model-independent, in the sense that it is the same relation that one would obtain by coupling linear realizations to supergravity and then, breaking spontaneously supersymmetry through the minimization of a given supersymmetric scalar potential[5)].

In the present contribution we extend this analysis to the case of extended supersymmetry with or without central charges. When central charges are absent, the extension of the Volkov-Akulov formalism is trivial. However, the more interesting situation of matter with non-trivial central charges[6)], or with central charges spontaneously broken is not trivial, and is the main result reported here. When non-linear Lagrangians with spontaneously broken central charges are coupled to N-extended supergravity, the simultaneous occurrence of the Higgs and superHiggs effect takes place. As a consequence, one obtains model-independent mass relations between the gravitinos, the gauge bosons of the supergravity multiplets, in terms of the vacuum energy and the Goldstone boson-goldstino-goldstino couplings of the global system. These couplings appear, in a purely geometrical fashion, in the non-linear globally supersymmetric original Lagrangian which describes the interaction of the Goldstone boson of central charges with the N-goldstinos.

We recall that the most general form of the N-extended supersymmetry algebra is[6)]

$$\{Q^i_\alpha, \bar{Q}^j_\beta\} = -2(\gamma^\mu)_{\alpha\beta} P_\mu \delta^{ij} - 2Z^{ij} 1_{\alpha\beta} - 2(\gamma_5)_{\alpha\beta} Z'^{ij} \quad (1)$$

0094-243X/84/1160067-06 $3.00

where Q^i_α are a set of N-translationally invariant spin-1/2 Hermitian operators (antisymmetric in the indices i,j) belonging to the centre of the superPoincaré algebra.

The group composition, in the infinitesimal, implies the following commutation relation

$$\left[\bar{\varepsilon}_1^i Q^i_\alpha, \bar{\varepsilon}_2^j Q^j_\alpha\right] = -2\bar{\varepsilon}_1^i \gamma^\mu \varepsilon_2^i P_\mu - 2\bar{\varepsilon}_1^i \varepsilon_2^j Z^{ij} - 2\bar{\varepsilon}_1^i \gamma_5 \varepsilon_2^j Z'^{ij} \tag{2}$$

where ε_1, ε_2 are spin-1/2 anticommuting constant spinors, and are interpreted as the parameter of a supersymmetric transformation.

If we denote now by $S_N IO(3,1)$ the N-extended superPoincaré algebra, an element of the coset-space $S_N IO(3,1)/O(3,1)$ is given by

$$\exp \frac{1}{d}\left[\bar{\theta} Q - iC_{ij} Z_{ij} - iC'_{ij} Z'_{ij}\right] \exp(-ix\cdot \mathrm{P}) \tag{3}$$

where d is a dimensionful constant, dim d = 2.

In the spirit of non-linear realizations the parameters θ_i, C_{ij}, C'_{ij} must be considered as fields, functions of space-time co-ordinates.

According to Callan, Coleman, Wess, Zumino and Weinberg[1)] the non-linear realizations of N-extended supersymmetry as well as of the N(N-1) central charges are obtained by the action of a group element characterized by parameters ε_i, ω_{ij}, ω'_{ij} on the coset-space given by Eq. (3).

This action induces a motion on the coset-space which is easily seen, by use of Eq. (2), to be of the form

$$\delta x^\mu = \frac{i}{d}\left[\bar{\varepsilon}^i \gamma^\mu \theta^i\right] = \xi^\mu \tag{4}$$

$$\delta \theta_i = d\varepsilon_i + \xi^\nu \partial_\nu \theta_i \tag{5}$$

$$\delta C_{ij} = d\omega_{ij} + \frac{i}{4}(\bar{\varepsilon}_i \theta_j - \bar{\varepsilon}_j \theta_i) + \xi^\nu \partial_\nu C_{ij} \tag{6}$$

$$\delta C'_{ij} = d\omega'_{ij} + \frac{i}{4}(\bar{\varepsilon}_i \gamma_5 \theta_j - \bar{\varepsilon}_j \gamma_5 \theta_i) + \xi^\nu \partial_\nu C'_{ij} \tag{7}$$

where

$$\xi^\mu = \frac{i}{d}\bar{\varepsilon}^i \gamma^\mu \theta^i$$

In order to construct Lagrangians which are invariant under the non-linear transformations (4)-(7), we determine the covariant derivatives of the fields introduced above.

Consider an element of the graded Lie algebra of the form

$$g^{-1}\delta g = \delta x^{\mu}\left\{-iV_{\mu}^{a}P_{a} + \frac{i}{d}C_{\mu ij}Z_{ij} + \frac{i}{d}C'_{\mu ij}Z'_{ij} + \frac{1}{d}\bar{g}_{\mu}^{i}Q^{i}\right\} \tag{8}$$

where g is a group element. A simple explicit computation gives

$$V_{\mu}^{a} = \delta_{\mu}^{a} + \frac{i}{d^{2}}(\bar{\theta}_{i}\gamma^{a}\partial_{\mu}\theta_{i}) \tag{9}$$

$$C_{\mu ij} = \partial_{\mu}C_{ij} - \frac{i}{4d}[\bar{\theta}_{i}\partial_{\mu}\theta_{j} - \bar{\theta}_{j}\partial_{\mu}\theta_{i}] \tag{10}$$

$$C'_{\mu ij} = \partial_{\mu}C'_{ij} - \frac{i}{4d}[\bar{\theta}_{i}\gamma_{5}\partial_{\mu}\theta_{j} - \bar{\theta}_{j}\gamma_{5}\partial_{\mu}\theta_{i}] \tag{11}$$

$$g_{\mu i} = \partial_{\mu}\theta_{i} \tag{12}$$

By construction, the above quantities behave, under a supersymmetry transformation, as covariant vectors:

$$\delta V_{\mu}^{a} = \xi^{\nu}\partial_{\nu}(V_{\mu}^{a}) + (\partial_{\mu}\xi^{\nu})V_{\nu}^{a} \tag{13}$$

The matrix $V_{\mu}^{a}(\theta)$ can be regarded as a covariant vierbein and its inverse $[V_{\mu}^{a}(\theta)]^{-1} = V_{a}^{\mu}(\theta)$ can be used to transform world-vectors into world scalars

$$C_{aij} = V_{a}^{\mu}C_{\mu ij} \tag{14}$$

$$\delta(C_{aij}) = \xi^{\nu}\partial_{\nu}(C_{aij}) \tag{15}$$

A general invariant action for the geometrical fields θ, C, C' has the form

$$A = \int d^{4}x(\det V) \qquad (C_{a}, C'_{a}, g_{a}) \tag{16}$$

where $\mathscr{L}$ is a Poincaré invariant function. The action given by Eq. (6) describes the most general interactions of the N-Goldstone fermion with N(N-1) Goldstone boson, associated with the spontaneous breaking of the corresponding central charges. A proper normalization of the Goldstone boson fields is

$$C_{ij} = \frac{1}{g_{ij}} Z_{ij} \ , \ C'_{ij} = \frac{1}{g'_{ij}} Z'_{ij} \tag{17}$$

The minimal Lagrangian is therefore

$$\mathscr{L}_{MIN} = -\frac{d^2}{2} \det V - \frac{1}{4} \det V \eta^{ab} \sum_{ij} [g_{ij}^2 C_{aij} C_{bij} + g'_{ij} 2 C'_{aij} C'_{bij}] \tag{18}$$

where

$$g_{ij} C_{\mu ij} = \partial_\mu Z_{ij} - \frac{i}{4} \frac{g_{ij}}{d} (\overline{\Theta}_i \partial_\mu \Theta_j - \overline{\Theta}_j \partial_\mu \Theta_i) \tag{19}$$

The coefficients have to be chosen so that θ_i, z_{ij}, z'_{ij} are property normalized fields.

From Eq. (19), it is obvious that when we set some of the $g_{ij} = 0$, the corresponding central charge is unbroken and the corresponding Z_{ij} becomes a normal neutral matter field which transforms non-linearly under supersymmetry but which is neutral under central charge transformations.

Irrespective of the fact whether g_{ij} vanish or not, we can introduce complex scalar fields non-neutral under central charge symmetry, so that

$$[Z_{ij}, \varphi] = i\sqrt{d} \ q_{ij} \varphi \qquad q_{ij} = -q_{ji} \tag{20}$$

Under supersymmetry ϕ transforms non-linearly

$$\delta\varphi = \xi^\mu \partial_\mu \varphi + i q_{ij} \xi_{ij} \varphi \tag{21}$$

and

$$\xi_{ij} = \frac{-i}{2\sqrt{d}} (\overline{\varepsilon}_i \Theta_j - \overline{\varepsilon}_j \Theta_i)$$

We can also introduce the covariant derivative of ϕ

$$D_\mu \varphi = \partial_\mu \varphi - \frac{1}{d\sqrt{d}} q_{ij} (\overline{\Theta}_i \partial_\mu \Theta_j) \varphi \tag{22}$$

so that

$$D_a \varphi = V_a^\mu D_\mu \varphi \tag{23}$$

transforms like Φ.

As a consequence the action

$$A_M = \int d^4x \, \det V \; \mathscr{L}_{MATTER} (\varphi, \varphi^*, D_a\varphi, D_a\varphi^*) \tag{24}$$

is invariant if $\mathscr{L}_{MATTER}$ is Poincaré invariant as well as invariant under U(1)-central charge transformation.

Of course, if some g_{ij} are non-vanishing, then the corresponding central charges are spontaneously broken, and one can write arbitrarily couplings provided ϕ is coupled to the Goldstone boson field as follows:

$$\psi = \exp\left[\frac{2i}{\sqrt{d}} q_{ij} \, C_{ij}\right]\varphi \tag{25}$$

so that ψ is inert under the action of Z_{ij}.

The coupling of non-linear realizations of supersymmetry and central charges to N-extended supergravity provides the simultaneous occurrence of the Higgs and superHiggs effect.

In order to work out the mass formulae which follow, it is sufficient to work out the full bilinear part of the coupled locally supersymmetric invariant Lagrangian as well as those trilinear couplings which are needed to make the bilinear couplings invariant. Of course, we are assuming that the full non-linear coupling exists. This may not be true for higher N theories. These results are derived in Ref. 7).

The cancellation of the cosmological constant implies the relation

$$m_{3/2}^2 = \frac{1}{6} k^2 d^2 \tag{26}$$

where k is the gravitational coupling constant.

The Higgs effect gives masses to the N(N-1)/2 vector bosons of the supergravity multiplet in terms of g_{ij} and $m_{3/2}$:

$$m_{ij}^2 = \frac{3}{16} g_{ij}^2 \, m_{3/2}^2 \tag{27}$$

The charge coupling of the vector bosons to the gravitino is

$$g_c = k \, m_{3/2} \tag{28}$$

and is universal.

Strictly speaking, these results, obtained by working out the linearized coupling, are only valid if $N < 4$. For $N \geq 4$ scalar fields appear in the supergravity multiplet and their v.e.v. can destroy the above relations. These relations remain valid only if extremes of the potential exist for which all scalar fields of the gravitational multiplet have vanishing v.e.v.s.

ACKNOWLEDGEMENTS

The material discussed here is a result of a collaboration with Prof. L. Maiani and Dr. P.C. West [see Ref. 7)].

REFERENCES

1. S. Weinberg, Phys. Rev. Lett. 18 (1967) 188;
S. Coleman, J. Wess and B. Zumino, Phys. Rev. 177 (1969) 2239;
C. Callan, S. Coleman, J. Wess and B. Zumino, Phys. Rev. 177 (1969) 2247.

2. V. Akulov and B. Volkov, Phys. Lett. 46B (1973) 109.

3. E. Ivanov and A. Kapustinkov, J. Phys. A11 (1978) 2375; J. Phys. G8 (1982) 167;
T. Uematsu and Z. Zachos, Nucl. Phys. B201 (1982) 250;
J. Wess, Karlsruhe preprint (1982).

4. S. Deser and B. Zumino, Phys. Rev. Lett. 38 (1977) 1433.

5. E. Cremmer et al., Phys. Lett. 79B (1978) 231; Nucl. Phys. B147 (1979) 105;
E. Cremmer, S. Ferrara, L. Girardello and A. van Proeyen, Phys. Lett. 116B (1982) 231; Nucl. Phys. B. B212 (1983) 413.

6. R. Haag, J.T. Lopuszanski and M. Sohnius, Nucl. Phys. B88 (1975) 257.

7. S. Ferrara, L. Maiani and P.C. West, CERN preprint TH.3513 (1983).

TWO THOUGHTS ON FLAVOR

Howard Georgi

Lyman Laboratory of Physics
Harvard University
Cambridge, MA 02138

I presume that I am here for comic relief. I am not going to talk about either unification or supergravity. Instead, I am going to discuss two ideas related to the flavor question. The first, TECHNIGIM, is the result of a collaboration which developed over large spatial and temporal distances with S. Dimopoulos and S. Raby.* The second, on the degeneracy of all fermions, is an idea of mine which I am presently working out with two students, Aneesh Manohar and Ann Nelson.

I. TECHNIGIM

Technicolor (TC) is a very attractive idea, which might address several of the fundamental puzzles of modern particle theory, but for the fact that it seems to lead to a phenomenological disaster. The SU(2) x U(1) breaking induced by TC must be communicated to the light quarks and leptons by some interaction which breaks the global chiral flavor symmetries. Fundamental scalars can do this communication, but such scalars are no more attractive than fundamental Higgs mesons. The only other possibilities would seem to be extended technicolor (ETC) gauge interactions which cause transitions between light fermions and technifermions or a dynamical model in which light fermions and technifermions are both built out of the same constituents. To date, all such schemes have been plagued by flavor changing neutral current (FCNC)

In this talk, I suggest a solution to the FCNC problem in TC theories. I will describe the solution in the language of ETC and then exemplify it in a specific ETC model. However, I believe that our mechanism is more general and applies to constituent models as well.

The problem with ETC theories is associated with the structure of an ETC gauge group. The ETC gauge generators and the corresponding gauge bosons are of three types: flavor (F) symmetry generators associated with transitions between flavors; TC generators; and generators associated with transitions from ordinary fermions to technifermions. This last type is required to generate light fermion masses. Henceforth, we will reserve the name ETC for these transition generators and gauge bosons. The flavor and TC generators must exist because they are produced by commutation of ETC generators with their adjoints. It is the flavor generators that cause the trouble. Typically, flavor gauge boson exchange contributes to $\Delta s = 2$ or $\Delta C = 2$ processes, or to both. These effects cannot be suppressed by simply increasing the flavor gauge boson

*Mediated by the Dimopoulon, which produces a force of infinite range.

0094-243X/84/1160073-04 $3.00

masses, because anything which increases the flavor gauge boson masses also increases the ETC gauge boson masses, which in turn decreases the light fermion masses.

To evade this snare, I turn to a generalization of the Glashow-Iliopoulos-Maiani (GIM) mechanism which banished FCNC effects from the standard model. In the standard model, there are no FCNC effects in lowest order because the gauge interactions have a very large flavor symmetry, SU(n) x SU(n) x SU(n) [x irrelevant U(1)'s] for n flavors. This flavor symmetry allows us to move the mixing angles in the quark sector from the charge 2/3 (U) quarks to the charge -1/3 (D) quarks or back at our convenience. FCNC effects can appear only when both U and D quarks are involved, as in the usual box diagram.

It is clear that this flavor symmetry argument cannot be trivially generalized to ETC. For one thin, in ETC theories, there are more gauge interactions. For the flavor interactions to have a flavor symmetry, the flavor gauge bosons must be degenerate. What is worse, the quark mass matrix cannot have the flavor symmetry, because the quarks are not degenerate. But the quark mass matrix comes from the gauge interactions. Thus the gauge interactions cannot have the flavor symmetry either.

The solution is simple:

Break the flavor symmetry where there is no mixing.
Introduce mixing only where there is flavor symmetry.

I will illustrate this mechanism in a toy model of quarks, in which the ETC group is a semisimple group, $SU(N)_L \times SU(N)_U \times SU(N)_D$. We ignore leptons and put in the ETC breaking by hand with fundamental scalar fields, in order to simplify the discussion and concentrate on the TECHNIGIM mechanism.

LH SU(2) doublets of quarks and techniquarks transform like Ns of $SU(N)_L$.

RH singlets of charge 2/3 (-1/3) quarks and techniquarks transform like N's of $SU(N)_U$ ($SU(N)_D$).

Introduce symmetry breaking which treats the L ETC very differently from the U and D ETC. The $SU(N)_L$ is broken directly down to SU(N−3), preserving an SU(3) global flavor symmetry of the $SU(N)_L$ flavor interactions. $SU(N)_U$ and $SU(N)_D$ are broken down to SU(N−3) in stages, preserving global U(1) symmetries but completely breaking the nonabelian flavor symmetries.

Finally, couple the various ETC's together. Break the three independent TCs down to a single diagonal TC. This produces a TC interaction which breaks SU(2) x U(1), but it is not enough to generate quark masses.

All of the above symmetry breaking scales are of the order of a few hundred GeV or larger. Introduce mixing between the flavor subgroups of $SU(N)_L$ and $SU(N)_{U \text{ and } D}$ at a lower scale μ. This produces quark masses proportional to μ. At this point, introduce nontrivial flavor mixing by inputting different mixings for L with U and L with D. For the simplest form of the mixing, to lowest order in μ, only the LH quarks get mixed. Then because of the SU(3) flavor symmetry of the LH gauge interactions, there is a GIM mechanism which eliminates FCNCs in lowest order in μ. Just as in the standard model, the mixing can be moved from U_L to D_L without changing the gauge interactions. The FCNCs from processes which involve both Us and Ds are suppressed by extra powers of the small scale μ. If μ (which sets the scale of all the quark masses) is small enough, the FCNCs will not cause phenomenological problems. This leads to a bound on the t quark mass,

$$m_t \leq 20\ GeV \cdot \left(\frac{1\ GeV}{m_c}\right)^{1/3} \cdot \left(\frac{\Lambda}{350\ GeV}\right)^2 .$$

This is, perhaps, just barely phenomenologically acceptable.

These simple ideas can be easily realized in a model in which the ETC gauge symmetries are broken by VEVs of fundamental scalar bosons. Details (and references) can be found in S. Dimopoulos, H. Georgi and S. Raby, Harvard Preprint HUTP-83-A002. We hope that the required symmetry breaking can itself be dynamically induced, but we have not yet constructed an explicit model.

It would be overstating the case to claim that this model is beautiful. But it works, for light quarks. Can we include leptons? Not without further enlarging the ETC group. I haven't eliminated the flavor interactions. If leptons are included in the same ETC groups as the quarks, we get a very large $K \to \mu e$ decay. Thus we need still more factors in our ETC gauge group. Explicit models will be discussed elsewhere.

More interesting, it seems to me, is the possibility that a mechanism like ours could arise dynamically in a subconstituent model. In a model of this kind, both quarks and techniquarks are built out of the same subconstituents, and they are light because the dynamics of the binding forces leaves some chiral symmetries unbroken. These are in turn broken by weaker gauge interactions which produce the analogs of the ETC interactions. Because our mechanism is essentially group theoretical, involving the flavor symmetry properties of the ETC and flavor interactions, we can hope to find a dynamical model with the same structure. Perhaps, in this way, we can avoid unwanted inflation of the ETC group.

II. ON THE DEGENERACY OF ALL FERMIONS

I now want to discuss an idea which could, I hope, lead to the calculation of quark and lepton masses. The idea is very simple, the families are unified in such a way that at the unification scale, all fermions are automatically degenerate. All fermions! Thus all mass ratios are renormalization effects, like m_b / m_τ in SU(5). But you way, Georgi has popped his cork. m_t / m_e is something like 10^5. How could any reliable perturbative calculation give such a large mass ratio? Here I will show you three things.

1. This is a simple class of models in which all the fermions are automatically degenerate at a large scale.

2. It **is** in fact possible to generate large mass ratios through perturbative renormalization.
3. The quark and lepton masses in such a theory have several features which are reminiscent of the observed masses. Alas, I will not discuss a specific model. I haven't found one that works, in the sense of being consistent with everything we know about the masses.

The only way I know to build realistic models in which all the fermions are automatically degenerate, independent of the form of VEVs, is to have the gauge group include an $SU(2)_L \times SU(2)_R$. Let G be the subgroup of the gauge group which commutes with $SU(2)_L \times SU(2)_R$. Then let the LH fermions transform under $SU(2)_L \times SU(2)_R \times G$ as $(2,1,R) + (1,2,\bar{R})$ where R is a complex irreducible representation of G. Then let the only $SU(2)_L$ doublet (and color SU(3) signlet) Higgs particles transform like the real components of a (2.2.1) under $SU(2)_L \times SU(2)_G \times G$. Then like the nucleons in the sigma model, all the fermions are degenerate if G is unbroken. Note that all these conditions are satisfied in GUTs based on large orthogonal groups with fermions in a spinor representation and the Higgs responsible for $SU(2)_L$ breaking fermion masses in a vector.

When the $SU(2)_R \times G$ symmetry is broken at a large scale M_G, the fermion representation breaks up into pieces which can transform differently under the remaining gauge symmetry. Then the Yukawa couplings of the Higgs doublet to the various quarks and leptons can be renormalized differently. Of course this does not happen if $SU(2)_L \times SU(2)_R \times G$ is broken at M_G all the way down to $SU(2)_L \times U(1) \times SU(3)$. We must preserve a larger gauge symmetry under which the various families transform differently.

Suppose $SU(2)_R \times G$ breaks down to $U(1) \times H$ where H is some semisimple group which contains the color SU(3). Under H, the representation R breaks up into pieces

$$R = \sum_{\oplus} r_i \ ,$$

with the different families in different H representations, r_i. Then the family masses are renormalized differently. But how large can these renormalizations be?

The renormalization of m_b / m_τ in SU(5) comes mostly from the color interaction. It is not very large in the standard model with three families, because the bulk of the contribution comes from scales near m_b, where the SU(3) coupling constant is not very small. Because of asymptotic freedom, the effect from larger scales is small. On the other hand, if there were significantly more than 16 quarks, so that the SU(3) coupling were rapidly increasing with energy scale,

GRAVITATIONALLY INDUCED BARYON DECAY

John Ellis and John S. Hagelin
Stanford Linear Accelerator Center
Stanford University, Stanford, California 94305

and

D. V. Nanopoulos
Theory Division, CERN
CH-1211 Geneva 23, Switzerland

and

K. Tamvakis
Department of Physics
University of Ioannina, Ioannina, Greece

ABSTRACT

We find that in supersymmetric theories gravitodynamic effects scaled by the inverse of the Planck mass can induce baryon decay at an observable rate. In a minimal supersymmetric (susy) grand unified theory (GUT) the dominant gravitationally induced baryon decay mode is $B \to \bar{\nu}+K$, with a likely admixture of $p \to (e^+ \text{ or } \mu^+)+K$. As a by-product, we present an improved estimate of Higgs-mediated baryon decay branching ratios in minimal susy GUTs. We consider the possibility that a loss of quantum coherence may be observable in gravitationally induced baryon decay, but argue that this would be difficult to reconcile with successful experimental tests of quantum mechanics.

Many theoretical frameworks for baryon decay have been discussed in the last few years. These include conventional grand unified theories (GUTs)[1], variants of supersymmetric GUTs[2−5], non-perturbative effects[6] in the electroweak gauge theory, and baryon number violation catalyzed by grand unified monopoles.[7] Zel'dovich[8] and others[9,10] have suggested that non-perturbative quantum gravitational effects such as virtual black holes may also lead to baryon decay, albeit with an unobservably long lifetime of order 10^{50} years or more.[10] Three of us[11] have recently pointed out that in a supersymmetric theory such a "gravitodynamic" baryon decay amplitude may be scaled by $O(1/m_P)$, leading to faster baryon decay which could be observable in the present generation of experiments, even if the putative grand unification mass is much larger than the canonical $O(10^{16})\ GeV$ encountered in minimal susy GUTs.

In this paper we explore the phenomenology of gravitodynamic baryon decay, assuming a susy GUT framework. We show that only a few distinct dimension 5 operators can be important, and quantify the baryon decay branching ratios to be expected in each case. We find that $B \to \bar{\nu} + K$ dominates, with a likely admixture of $p \to (e^+ \ or\ \mu^+) + K$. As a by-product, we correct previous estimates[2,3,12,13] of baryon decay branching ratios in conventional susy GUTs, confirming the dominance[3] of $B \to \bar{\nu} + K$ induced by $\tilde{W}$ exchange. We show that the baryon decay rate can be observable even if the dimensional coefficient of the gravitationally induced dimension 5 operators is much smaller than $m_P^{-1} = O(10^{-19})\ GeV^{-1}$. We discuss the possibility[9,10] that quantum coherence may be lost in baryon decay and other processes induced by non-perturbative quantum gravitational phenomena such as virtual black holes.

We assume that particle physics is described by a renormalizable spontaneously broken local susy gauge theory at energies much below the Planck mass, augmented by non-renormalizable interactions scaled by the appropriate inverse power of the Planck mass.[11] We allow these non-renormalizable interactions to violate all global symmetries consistent with gauge invariance and susy. The requirement of susy is essential in order to protect the squarks $\tilde{q}$ and sleptons $\tilde{\ell}$ from acquiring masses $O(m_P)$, and also to prevent the appearance of quartic scalar interactions such as $\tilde{q}\tilde{q}\tilde{q}\tilde{\ell}$ with coefficients of order unity, which would lead to catastrophically rapid baryon decay. A general effective phenomenological action framework for these interactions is provided by the work of Cremmer *et al.*[14] We expect such interactions by analogy with low energy effective interactions in strong interaction physics, and are agnostic about their detailed origins.

Maybe there are many particles with masses $O(m_P)$ which can mediate new interactions. Or maybe our known "elementary" particles are in fact composite at the Planck scale,[15] with novel interactions corresponding to constituent interchanges. Another possible source is non-perturbative quantum gravitational effects such as virtual black holes[8] or space-time foam.[9,10] * We would certainly expect an effective Lagrangian approach to be applicable in the first two cases, but it has been suggested[9] that in the latter case quantum coherence might be lost, entailing a revision[10] of conventional quantum field theoretical rules. Later on, however, we will encounter reasons for discounting the observability of incoherence, and for playing according to the field-theoretical rules, at least as a first approximation.

We will be concerned with possible non-renormalizable terms in the chiral superpotential, and the lowest-dimensional terms of interest are quartic. We assume an SU(5) GUT framework so that quarks and leptons appear in chiral $\overline{5}$ ($\overline{F}^{\alpha}$) and $\underline{10}$ ($T_{\beta\gamma}$) superfields. The only interesting quartic combinations which can violate B and L conservation are

$$\overline{F}^{\alpha}_{a}\, T^{b}_{\alpha\beta} T^{c}_{\gamma\delta} T^{d}_{\lambda\mu}\ \epsilon^{\beta\gamma\delta\lambda\mu} \tag{1}$$

where the Latin indices denote different generations of quarks and leptons. These give the same dimension 5 operators[16] as Higgs exchange in minimal susy GUTs,[17] but in a potentially different algebraic combination. One can extract from (1) $\Delta B = \Delta L = \pm 1$ interactions of the form $\ell^{a}_{L} q^{b}_{L} q^{c}_{L} q^{d}_{L}$ and $q^{c^a}_{L}\, \ell^{c^b}_{L}\, q^{c^c}_{L}\, q^{c^d}_{L}$. Because of colour antisymmetrization, the three quark generation indices cannot all be the same, and none of the $q^{c}_{L} \ell^{c}_{L} q^{c}_{L} q^{c}_{L}$ operators can contribute to baryon decay when dressed by gaugino exchange. However the following $\ell_L q_L q_L q_L$ operators can contribute when dressed by gaugino exchange:

* We note that extant foam[9,10] calculations respect discrete reflection symmetries on the matter fields: $\phi \to -\phi$, and we assume such a symmetry in the following. This has the effect of forbidding catastrophic B and L-violating trilinear qqq and $q\,\bar{q}\,\ell$ interactions.

$$\mathcal{L}_{eff} = \int d^2\theta\epsilon^{ijk} \sum_{\ell=e,\mu,\tau} \Big[g^{\ell}_{cuu}(\ell c_i u_j d'_k - \nu_\ell s'_i u_j d'_k) + g^{\ell}_{ucc}(\ell u_i c_j s'_k - \nu_\ell d'_i c_j s'_k)$$

$$+ g^{\ell}_{tuu}(\ell t_i u_j d'_k - \nu_\ell b'_i u_j d'_k) + g^{\ell}_{utt}(\ell u_i t_j b'_k - \nu_\ell d'_i t_j b'_k)$$

$$+ \frac{1}{2} g^{\ell}_{uct}(\ell u_i c_j b'_k - \ell u_i s'_j t_k - \nu_\ell d'_i c_j b'_k + \nu_\ell d'_i s'_j t_k)$$

$$+ \frac{1}{2} g'^{\ell}_{uct}(\ell c_i t_j d'_k - \ell c_i b'_j u_k - \nu_\ell s'_i t_j d'_k + \nu_\ell s'_i b'_j u_k) \Big] . \tag{2}$$

where the indices i,j,k denote SU(3) colour and the Cabibbo-rotated charge -1/3 quarks are denoted by primes. We might naively expect on dimensional grounds that the coefficients g in (2) would be $O(m_P^{-1}) = O(10^{-19})\, GeV^{-1}$. The superspace integration $\int d^2\theta$ picks out pairs of left-handed fermion components from each of the products of four superfields in the sum (2). The resulting two fermion-two boson products can then be dressed by SU(2) $\tilde{W}^{\pm,0}$, U(1) $\tilde{B}$ or SU(3) $\tilde{g}$ gaugino exchange to give four-fermion operators.[16] The flavour-conserving $\tilde{B}$ and $\tilde{g}$ exchanges cannot possibly give operators relevant to baryon decay, except from the terms $\epsilon^{ijk}\nu_\ell s'_i u_j d'_k$, $\epsilon^{ijk}\nu_\ell b'_i u_j d'_k$ and $\epsilon^{ijk}\nu_\ell s'_i b'_j u_k$. However, $\tilde{B}$ and $\tilde{g}$ exchanges both generate from these operators symmetric combinations of four-fermion operators such as*

$$\epsilon^{ijk} \left[(\nu_\ell s'_i)_L (u_j d'_k)_L + (\nu_\ell d'_k)_L (s'_i u_j)_L + (\nu_\ell u_j)_L (s'_i d'_k)_L \right] \tag{3}$$

where $(ff')_L$ denotes the Lorentz scalar product of two left-handed fermion fields. However, this combination vanishes because of a simple algebraic identity.** Therefore $\tilde{B}$ and $\tilde{g}$ dressings of dimension 5 operators do not contribute to baryon decay. However, dressing the dimension 5 operators with $\tilde{W}^{\pm}$ gauginos can contribute to baryon decay.

---

*This point was overlooked by Aliev and Vysotsky in Ref. 13.

**This cancellation was missed in Ref. 2.

Exchanges of the complete isotriplet $\tilde{W}^{\pm,0}$ gives rise to the following unrenormalized $SU(3) \times SU(2) \times U(1)$ invariant four-fermion operators (in the notation of Ref. 3):

$$\mathcal{L}^0_{eff} = b^0 \sum_{\ell=e,\mu,\tau} \left[g^{\ell}_{cuu} \left\{ \frac{1}{2} [O^4_{cuu\ell} - 3O^3_{cuu\ell} + 3O^3_{uuc\ell}] \right\} \right.$$

$$+ g^{\ell}_{ucc} \left\{ \frac{1}{2} [O^4_{ucc\ell} - 3O^3_{ucc\ell} + 3O^3_{ccu\ell}] \right\} \tag{4}$$

$$\left. + \, similar \; terms \right] .$$

where b^0 is a loop integral:

$$b^0 = \frac{G_F}{\sqrt{2}} \frac{m^2_W m_{\tilde{W}}}{32\pi^2} [f(m_{\tilde{q}}, m_{\tilde{q}}, m_{\tilde{W}}) + f(m_{\tilde{q}}, m_{\tilde{\ell}}, m_{\tilde{W}})] \tag{5a}$$

with

$$f(m_1, m_2, m_3) \equiv \frac{1}{m^2_2 - m^2_3} \left(\frac{m^2_2}{m^2_1 - m^2_2} \, \ell n \frac{m^2_1}{m^2_2} - \frac{m^2_3}{m^2_1 - m^2_3} \, \ell n \frac{m^2_1}{m^2_3} \right) . \tag{5b}$$

If we guess

$$\left(\frac{m_{\tilde{W}}}{m_W} \right) [f(m_{\tilde{q}}, m_{\tilde{q}}, m_{\tilde{W}}) + f(m_{\tilde{q}}, m_{\tilde{\ell}}, m_{\tilde{W}})] \simeq \frac{1}{m^2_W} \tag{6}$$

we get

$$b^0 \simeq 2.2 \times 10^{-6} \; GeV^{-1} . \tag{7}$$

This result is renormalized in the usual way by gauge loop corrections. Ignoring renormalization between distances m_P^{-1} and m_X^{-1}, the renormalization at larger distance scales is

$$A = \left[\frac{\alpha_3(1GeV)}{\alpha_3(m_c)} \right]^{2/9} \left[\frac{\alpha_3(m_c)}{\alpha_3(m_b)} \right]^{6/25} \left[\frac{\alpha_3(m_b)}{\alpha_3(m_t)} \right]^{6/23} \left[\frac{\alpha_3(m_t)}{\alpha_3(m_W)} \right]^{2/7}$$

$$\left[\frac{\alpha_3(m_W)}{\alpha_{SUM}} \right]^{4/3} \left[\frac{\alpha_2(m_W)}{\alpha_{SUM}} \right]^{-3} \left[\frac{\alpha_1(m_W)}{\alpha_{SUM}} \right]^{-1/66} \tag{8}$$

which takes the value

$$A \simeq 15 \tag{9}$$

when we make the illustrative choices

$$\alpha_3(m_W) = 0.12, \alpha_2(m_W) = \frac{1}{31}, \alpha_1(m_W) = \frac{1}{50}, \alpha_{SUM} = \frac{1}{24} \,. \tag{10}$$

The coefficient of the renormalized operators is therefore

$$b \;\equiv\; Ab^0 \simeq 3.2 \times 10^{-5}\, GeV^{-1} \,. \tag{11}$$

In order to see whether this susy gravitodynamic mechanism can lead to baryon decay at an observable rate, we exploit the non-relativistic SU(6) analysis by Salati and Wallet[12] of the baryon decay modes induced by a general $\mathcal{L}_{eff}$. In their notation we have

$$\begin{aligned} A^L(\overline{\nu}_R\,;\Delta S = 1) = b\Big[& g^{\ell}_{cuu}(3U_{cs}U_{ud} - U_{cd}U_{us}) - 2g^{\ell}_{ucc}U_{cs}U_{cd} \\ & + g^{\ell}_{tuu}(3U_{ts}U_{ud} - U_{td}U_{us}) - 2g^{\ell}_{utt}U_{ts}U_{td} \\ & - g^{\ell}_{uct}(U_{ts}U_{cd} + U_{td}U_{cs}) + \frac{1}{2}g'^{\ell}_{uct}(3U_{td}U_{cs} - U_{ts}U_{cd})\Big] \end{aligned} \tag{12a}$$

$$\begin{aligned} A'^L(\overline{\nu}_R\,;\Delta S = 1) = b\Big[& g^{\ell}_{cuu}(3U_{cd}U_{us} - U_{cs}U_{ud}) - 2g^{\ell}_{ucc}U_{cs}U_{cd} \\ & + g^{\ell}_{tuu}(3U_{td}U_{us} - U_{ts}U_{ud}) - 2g^{\ell}_{utt}U_{ts}U_{td} \\ & - g^{\ell}_{uct}(U_{ts}U_{cd} + U_{td}U_{cs}) + \frac{1}{2}g'^{\ell}_{uct}(3U_{ts}U_{cd} - U_{td}U_{cs})\Big] \end{aligned} \tag{12b}$$

$$\begin{aligned} B(\overline{\nu}_R\,;\Delta S = 1) = b\Big[& -g^{\ell}_{cuu}(U_{cs}U_{ud} - U_{cd}U_{us}) \\ & - g^{\ell}_{tuu}(U_{ts}U_{ud} - U_{td}U_{us}) \\ & + \frac{1}{2}g'^{\ell}_{uct}(U_{ts}U_{cd} - U_{td}U_{cs})\Big] \end{aligned} \tag{12c}$$

$$A^L(\overline{\nu}_R\,;\Delta S=0)=b\Big[2g^{\ell}_{cuu}U_{cd}U_{ud}-2g^{\ell}_{ucc}U^2_{cd}$$

$$+\,2g^{\ell}_{tuu}U_{td}U_{ud}-2g^{\ell}_{utt}U^2_{td} \qquad 12d)$$

$$-2g^{\ell}_{uct}U_{td}U_{cd}+g'^{\ell}_{uct}U_{td}U_{cd}\Big]$$

$$A^L(\ell^+_R;\Delta S=1)=b\Big[2g^{\ell}_{cuu}U_{cs}+2g^{\ell}_{tuu}U_{ts}\Big] \qquad (12e)$$

$$A^L(\ell^+_R;\Delta S=0)=b\Big[2g^{\ell}_{cuu}U_{cd}+2g^{\ell}_{tuu}U_{td}\Big] \qquad (12f)$$

where the $U_{qq'}$ are the appropriate entries in the Cabibbo-Kobayashi-Maskawa matrix which we assume[3] to be the same for squarks as for quarks. Evidently the $\Delta S=0$ decay modes are Cabibbo-suppressed relative to $\Delta S=1$ modes, irrespective of which of the g^{ℓ}_{abc} may dominate. For the Cabibbo-favoured decay modes $p,n\to\overline{\nu}+K,\overline{\nu}+K^*,e+K,\mu+K$ and $e+K^*$, the results of Salati and Wallet[12] give

$$\Gamma(p\to\overline{\nu}+K^+)=1.39\times10^{27}\cdot\left|A^L(\overline{\nu}_R\,;\Delta S=1)-B(\overline{\nu}_R\,;\Delta S=1)\right|^2y^{-1} \qquad (13a)$$

$$\Gamma(p\to\overline{\nu}+K^{*+})=1.23\times10^{25}\cdot\left|3A^L(\overline{\nu}_R\,;\Delta S=1)+2A'^L(\overline{\nu}_R\,;\Delta S=1)-B(\overline{\nu}_R\,;\Delta S=1)\right|^2y^{-1} \qquad (13b)$$

$$\Gamma(p\to e^++K^0)=1.39\times10^{27}\cdot\left|A^L(e^+_R;\Delta S=1)\right|^2y^{-1} \qquad (13c)$$

$$\Gamma(p\to\mu^++K^0)=1.34\times10^{27}\cdot\left|A^L(\mu^+_R;\Delta S=1)\right|^2y^{-1} \qquad (13d)$$

$$\Gamma(p \to e^+ + K^{*0}) = 1.23 \times 10^{25} \cdot \left|A^L(e^+_R; \Delta S = 1)\right|^2 y^{-1} \quad (13e)$$

$$\Gamma(n \to \overline{\nu} + K^0) = 1.37 \times 10^{27} \cdot \left|A^L(\overline{\nu}_R; \Delta S = 1) + A'^L(\overline{\nu}_R; \Delta S = 1)\right|^2 y^{-1} \quad (13f)$$

$$\Gamma(n \to \overline{\nu} + K^{*0}) = 1.28 \times 10^{25} \cdot \left|-3A^L(\overline{\nu}_R; \Delta S = 1) - A'^L(\overline{\nu}_R; \Delta S = 1) + 2B^L(\overline{\nu}_R; \Delta S = 1)\right|^2 y^{-1} \quad (13g)$$

Inserting the coefficients (12) into these formulae for the partial rates, we obtain the results of our Table showing the baryon decay branching ratios which follow from dominance by the different operators in $\mathcal{L}_{eff}$ (2).* In the absence of strong generation-dependence in the coefficients g, we would expect the operators g^{ℓ}_{cuu} to dominate, since they contribute to baryon decay without Cabibbo-Kobayashi-Maskawa suppression. As mentioned above, we see that $B \to \overline{\nu} + K$ dominates, with an admixture of $p \to \ell^+ + K$ if g^{ℓ}_{cuu} or g^{ℓ}_{tuu} are the dominant operators, and with relatively small branching ratios into $\overline{\nu} + K^*$. It is interesting to note that because the operators (12) do not all have definite strong isospin, the decay rates of the p and n into $\overline{\nu} + K^{+,0}$ (or $\overline{\nu} + K^{*+,0}$) differ in general.

It is interesting to note that a very different perspective on these relative decay rates is provided by a chiral Lagrangian formalism, according to which *none of our operators produce nucleon decay into pseudoscalar mesons.* By chiral arguments, or equivalently by 'soft pion' or 'soft kaon' theorems in current algebra, the matrix element of these $(\ell q)_L (qq)_L$ operators between a nucleon and pseudoscalar meson can be reduced to a matrix element of the form $\langle N | q_L q_L q_L | 0 \rangle$ which *vanishes.* This suggests that nonrelativistic SU(6) may overestimate the rates for (v or ℓ) + K in susy models (whether gravitodynamic or from conventional Higgs exchange), and suggests that branching ratios for vector mesons may be larger than in the Table.

---

* We note in passing that conventional minimal supersymmetric GUTs should have the same pattern of decay modes with $g^{\mu}_{ucc} : g^{e}_{ucc} : g^{\mu}_{cuu} : g^{e}_{cuu} = m_c m_s sin\theta : m_c m_d cos\theta : m_u m_s cos\theta : -m_u m_d sin\theta$ These results differ from those of Ref. 12 because the analogue of equation (4) was misprinted in Ref. 3.

TABLE I

Relative Decay Rates

Dominant Operator	$p \to \bar{\nu}K^+$	$p \to e^+K^o\,(\mu^+K^o)$	$p \to \bar{\nu}K^{*+}$	$p \to e^+K^{*o}$	$n \to \bar{\nu}K^o$	$n \to \bar{\nu}K^{*o}$
$g^{e(\mu)}_{cuu}$ (s^0), $g^{e(\mu)}_{tuu}$ (s^2)	1	$\frac{1}{4}$	$\mathcal{O}\left(\frac{1}{25}\right)$	$\mathcal{O}\left(\frac{1}{400}\right)$	$\frac{1}{4}$	$\mathcal{O}\left(\frac{1}{16}\right)$
$g^{e(\mu)}_{ucc}$ (s^2), $g^{e(\mu)}_{uct}$ (s^4), $g^{e(\mu)}_{utt}$ (s^6)	$\frac{1}{4}$	0	$\mathcal{O}\left(\frac{1}{16}\right)$	0	1	$\mathcal{O}\left(\frac{1}{25}\right)$
$g'^{e(\mu)}_{uct}$ (s^4)	$\mathcal{O}(1)$	0	$\mathcal{O}\left(\frac{1}{100}\right)$	0	$\mathcal{O}(1)$	$\mathcal{O}\left(\frac{1}{100}\right)$

The decay rates into K^* are suppressed by kinematic factors. The relative decay rates for the g'^{ℓ}_{uct} case depend on unknown Cabibbo-Kobayashi-Maskawa mixing angles. Indicated in parentheses at the left are the number of small angle factors suppressing the decay rates due to the different operators. Decay rates into non-strange final states are suppressed by an additional factor of $\sin^2\theta_c$.

Comparing the decay rates in (13) with the present lower limit[18] on the nucleon lifetime of order 3×10^{30} years, we infer that

$$g^{e,\mu}_{cuu} < O(10^{-6})\frac{1}{m_P} \tag{14}$$

and similarly for the other operator coefficients in equation (2). We do not know *a priori* what the values of these coefficients g might be. Perhaps they should be $O(1/m_P)$, or perhaps they are suppressed by analogy with the trilinear Yukawa couplings of chiral superfields. Equation (14) suggests that gravitodynamic baryon decay could even be overdue in susy theories, whereas it would be unobservably slow in theories without light spin-zero squark and slepton fields. It is perhaps surprising that the expected hierarchy of decay modes in the table is so similar to that coming[3] from conventional minimal susy GUTs, whereas one might have expected that no hard and fast predictions could be made about gravitationally induced baryon decay modes.

It has been suggested[9,10] that quantum coherence might be lost in gravitationally induced interactions, and it is reasonable to ask whether this could be detected in baryon decay, whether or not a satisfactory framework exists[10] for describing such a breakdown of quantum mechanics. A natural idea is to study the distribution of spins in baryon decay through an experiment analogous to those[19,20] inspired by Einstein, Podolsky and Rosen.[21] Such an experiment could in principle proceed in the following manner. We will consider for concreteness the case of $p \rightarrow \mu^+ + \rho$. Once the nonrelativistic ρ decays into $\pi\pi$, one can define the $\pi\pi$ axis to be the $\hat{z}$ axis. Since a ρ with spin $S_{\hat{z}} = \pm 1$ cannot give rise to final state mesons whose momenta lie exactly along the $\hat{z}$ axis, one can deduce that the ρ was precisely $S_{\hat{z}} = 0$. However, quantum mechanics is incompatible with the interpretation that the ρ was produced with any definite spin state. It is the decay (or more precisely the observability of the $\pi\pi$ decay products) which determines the spin state precisely. At this point, if the ρ and the muon are produced "coherently" (i.e., in a pure state) the muon also assumes a definite spin state (i.e., "collapses") such that angular momentum is conserved in the quantum mechanical sense. Thus, for example, if the initial baryon state polarization were known and one simply assumes that the baryon decays in an S-wave, one obtains a definite set of predictions for the muon's spin distribution. If, instead, the μ^+ and ρ are produced incoherently and behave afterwards as independent particles (as in thermal radiation from black hole evaporation), one would expect the muon spin wave function to collapse independently of the ρ, which could change the predicted spin distribution. Unfortunately, baryon decay modes such as $B \rightarrow \ell^+ +$ vector meson that have analyzable spin states are Cabibbo-and phase space-suppressed, and have branching ratios of order 1%. Moreover, the experiment requires polarized nucleons and the ability to measure the lepton spins, neither of which seems particularly feasible in a large scale baryon decay experiment. Furthermore, a change in spin distribution would entail a microscopic breakdown of angular momentum conservation in the quantum mechanical sense. A more palatable possibility might be the appearance of interactions which respect quantum mechanical gauge invariance and angular momentum conservation, but are mixed in flavor space. This would imply treating interactions of the type of eq. (1) *incoherently* in the generation indices, and hence discarding interferences between the various g^{ℓ}_{qqq} in eq. (12), instead of adding them coherently. This could lead to baryon decay branching ratios different from those obtained from eqs. (12) and (13) if two or more of the interactions have comparable magnitude.* Given the current stage-fright of unstable baryons, it is unlikely that the precision measurements required to discern such an effect will be forthcoming. However, in addition to these practical problems, we have the following theoretical reasons to believe that quantum mechanics would not be violated observably in baryon decay.

In the same way that one might expect interactions like (1) to be generated, one might also expect[11,22] to encounter interactions like

$$O\left(\frac{1}{m_P}\right)(F\overline{H}\Sigma\,T)\ ,\ O\left(\frac{1}{m_P}\right)(TT\Sigma\,H) \tag{15}$$

where H, $\overline{H}$ and Σ are respectively a $\underline{5}$, $\underline{\overline{5}}$ and $\underline{24}$ of Higgs in SU(5). Putting in $\langle 0|H,\overline{H}\,|0\rangle = 0(100)GeV$ and $\langle 0|\,\Sigma|0\rangle = O(10^{16})GeV$ one easily gets from (15) contributions to the e, d and u masses which are of the same order as their actual values.[11,22] However, we know that non-relativistic electrons and neutrons obey quantum mechanics with very high precision. Therefore there are severe upper limits on the amount of incoherence that can be introduced by gravitational effects analogous to mass terms arising from interactions such as (15). The best limit may come from experiments[23] on quantum interference using slow neutrons. Coherence could not be maintained over the times and distances used in these experimental tests of quantum mechanics unless

$$\frac{\delta m_{incoherent}}{m_n} \leq O(10^{-18}) \tag{16}$$

* It is possible that this discussion could be formalized in a useful way through a set of Bell-like inequalities. The postulate of local realism, for example, led to a set of predictions (known as Bell's inequalities) which any hidden variable theory must satisfy. Quantum mechanics violates these inequalities, since it is, in a sense, less deterministic than the postulate of local realism requires. The new violations of quantum mechanics suggested by Hawking[9,10] imply a higher level of unpredictability and hence indeterminacy than in conventional quantum mechanics. This suggests that a set of 'super-Bell' identities could be formulated which would be satisfied both by hidden variable theories and by quantum mechanics, but which these incoherent gravitodynamic effects would not satisfy.

which can be translated into

$$\delta m_q(incoherent) \leq O(10^{-18})GeV \ . \tag{17}$$

A similar, if somewhat less restrictive, constraint follows from the coherent behavior of kaons in $K^0 - \overline{K^0}$ mixing. We therefore deduce from (17) that gravitodynamic interactions such as (15) must be predominantly coherent if they are present at anywhere near the expected[22] level. One may expect the same to be true of the baryon decay operator (1): if it is observable it will probably be coherent. Note that although our remarks about quantum mechanics were motivated by the suggestion[9,10] that non-renormalizable interactions scaled by $O(1/m_P)$ might not be coherent, the restrictions (16) and (17) also constrain severely any incoherent contribution to conventional renormalizable interactions including Yukawa or gauge couplings. We should mention that some arguments in Ref. 10 suggest that some incoherent non-perturbative gravitational effects at low energies E may be suppressed by additional powers of $O(E/m_P) = O(10^{-19})$. * In this case, a modest improvement of the limit (16) by just a few orders of magnitude could be very revealing.

ACKNOWLEDGEMENTS

We would like to thank H. Georgi, V. Kaplunovsky, M. Perry, M. Peskin, J. Polchinski, M. Srednicki, N. Warner, and F. Wilczek for interesting comments.

REFERENCES

1. For reviews, see P. Langacker, Phys. Rep. *72C*, 185 (1981); Proceedings of the 1981 International Symposium on Lepton and Photon Interactions at High Energies, ed. W. Pfeil (Universität Bonn, 1981), p. 823.
2. S. Dimopoulos, S. Raby and F. Wilczek, Phys. Lett. *112B*, 133 (1982).
3. J. Ellis, D. V. Nanopoulos and S. Rudaz, Nucl. Phys. *B202*, 43 (1982).

---

* This suppression might be absent in a more realistic theory where the space-time foam is supersymmetric. Also, we know there are matter interactions violating global symmetries, which could for example generate $O(1/m_P)$ interactions via radiative corrections to the effects discussed in Ref. 10.

4. D. V. Nanopoulos and K. Tamvakis, Phys. Lett. *113B*, 151 (1982) and *114B*, 235 (1982).

5. A. Masiero, D. V. Nanopoulos, K. Tamvakis and T. Yanagida, Phys. Lett. *115B*, 298 (1982); Y. Igarashi, J. Kubo and S. Sakakibara, Phys. Lett. *116B*, 349 (1982).

6. G. 't Hooft, Phys. Rev. Lett *37*, 8 (1976).

7. V. A. Rubakov, Zh. Eksp. Teor. Fiz. Pis'ma Red. *33*, 658 (1981) [J.E.T.P. Lett. *33*, 644 (1981)]; Nucl. Phys. *B203*, 311 (1982); C. G. Callan, Phys. Rev. *D26*, 2058 (1982) and Princeton University preprint (1982); J. Ellis, D. V. Nanopoulos and K. A. Olive, Phys. Lett. *116B*, 127 (1982); F. A. Bais, J. Ellis, D. V. Nanopoulos and K. A. Olive, CERN preprint TH-3383 (1982).

8. Ya. B. Zel'dovich, Phys. Lett. *59A*, 254 (1976): Zh. Eksp. Teor. Fiz. *72*, 18 (1977) [Sov. Phys. J.E.T.P. *45*, 9 (1977)]; see also B. K. Harrison, K. S. Thorne, M. Wakano and J. S. Wheeler, "Gravitational Theory and Gravitational Collapse" (University of chicago Press, Chicago, 1965), p. 146.

9. S. W. Hawking, D. N. Page and C. N. Pope, Phys. Lett. *86B*, 175 (1979), Nucl. Phys. *B170*, 283 (1980).

10. S. W. Hawking, D.A.M.T.P. Cambridge preprint "The Unpredictability of Quantum Gravity" (May 1982).

11. J. Ellis, D. V. Nanopoulos and K. Tamvakis, CERN preprint TH-3418 (1982).

12. P. Salati and C. Wallet, Nucl. Phys. *B209*, 389 (1982); see also M. B. Gavela *et al.*, Phys. Rev. *D23*, 1580 (1980); N. Isgur and M. B. Wise, Phys. Lett. *117B*, 179 (1982) and erratum.

13. W. Lucha, University of Vienna preprint, WUThPh-1982-29 (1982); T. M. Aliev and M. I. Vysotsky, Phys. Lett. *120B*, 119 (1983).

14. E. Cremmer, B. Julia, J. Scherk, S. Ferrara, L. Girardello and P. van Nieuwenhuizen, Phys. Lett. *79B*, 231 (1978); Nucl. Phys. *B147*, 105 (1979). E. Cremmer, S. Ferrara, L. Girardello and A. Van Proeyen, CERN preprints TH-3312 and 3348 (1982).

15. J. Ellis, M. K. Gaillard, L. Maiani and B. Zumino, "Unification of the Fundamental Particle Interactions," ed. S. Ferrara, J. Ellis and P. van Nieuwenhuizen (Plenum

Press, N.Y., 1980), p. 69; J. Ellis, M. K. Gaillard and B. Zumino, Phys. Lett. *94B*, 343 (1980).

16. S. Weinberg, Phys. Rev. *D26*, 187 (1982); N. Sakai and T. Yanagida, Nucl. Phys. *B197*, 533 (1982).

17. S. Dimopoulos and H. Georgi, Nucl. Phys. *B193*, 150 (1981); N. Sakai, Zeit. für Phys. *C11*, 153 (1982).

18. M. R. Krishnaswamy *et al.*, Phys. Lett. *115B*, 349 (1982); G. Battistoni *et al.*, Phys. Lett. *118B*, 461 (1982).

19. J. S. Bell, Physics *1*, 195 (1964).

20. J. F. Clauser, M. A. Horne, A. Shimony and R. A. Holt, Phys. Rev. Lett. *23*, 880 (1969).

21. A. Einstein, B. Podolsky and N. Rosen, Phys. Rev. *47*, 777 (1935).

22. J. Ellis and M. K. Gaillard, Phys. Lett. *88B*, 315 (1979).

23. R. Colella, A. W. Overhauser and S. A. Werner, Phys. Rev. Lett. *34*, 1472 (1975); J.-L. Staudenmann, S. A. Werner, R. Colella and A. W. Overhauser, Phys. Rev. *21A*, 1419 (1980).

RADIATIVE SU(2) x U(1) BREAKING FROM N = 1 SUPERGRAVITY

L.E. Ibánez†
Departmento de Fisica Teórica
Universidad Autónoma de Madrid
Cantoblanco, Madrid-34, Spain

ABSTRACT

We consider the coupling of grand unified theories to N = 1 supergravity. The breaking of N = 1 local supersymmetry is only manifested at low energies through soft terms explicitly breaking the global supersymmetry. These soft terms (scalar masses, gaugino masses and trilinear scalar couplings) are renormalized at low energies according to the renormalization group. The $(\text{mass})^2$ of the Higgs doublet evolves towards negative values giving rise to SU(2) x U(1) breaking. This is achieved for masses of the top quark $m_t \gtrsim 55 GeV$. If gluino masses are $M_g \lesssim 100 GeV$, the breaking of SU(2) x U(1) requires $m_t \gtrsim 100\ GeV$.

Supersymmetry[1] seems to be a promising way to solve the gauge hierarchy problem. One may build specific supersymmetric GUT's (Susy-GUT's) in which susy is broken at a scale $\sim (\alpha)^{-1} M_w \sim 1\, TeV$, in order to maintain the W-S doublets massless down to the weak scale.[2-5] It has been recently realized[6-8] that, in fact, one may break supersymmetry at a very large mass scale, as long as one does it in such a way that the W-S doublets do not get large radiative masses after susy-breaking. This may be the case if one breaks susy in a very heavy ($\sim M_x$) and/or SU(5) singlet sector. Supersymmetry may then be broken at scales of order $10^{10} - 10^{19} GeV$.[6-8] If that is the case one probably cannot ignore the effects of supergravity. Since the observed low energy spectrum is chiral, an obvious possibility to study the effects of supergravity is to consider the coupling of N = 1 supergravity to GUT's. This has been recently studied by various authors[9-12] (e.g. the talks by Arnowitt and Weinberg in these proceedings). Let us briefly recall the most relevant formulae.

In superspace notation, the action for the coupling of chiral superfields φ and Yang-Mills superfields V to supergravity takes the form[13].

$$\int dx^4 d\vartheta^2 d\bar{\vartheta}^2 E \left\{ J\left[\bar{\varphi} e^{2gV}, \varphi\right] + \text{Re} \left[\frac{1}{R} W(\varphi) + \frac{1}{R} f_{ab}(\varphi) F^a F^b \right] \right\} \tag{1}$$

where E is the superspace determinant and R is the scalar curvature superfield. $W(\varphi)$ is the superpotential and F^a is the supersymmetric chiral field strength tensor (a is a gauge index). The functions J and f represent non-canonical modifications of the kinetic terms for chiral and vector superfields, respectively. It turns out that the final Lagrangian written in terms of the component fields only depends on J and W through the following combination:[13]

$$G(\varphi,\varphi^*) = 3\, Log \left(\frac{-J(\varphi,\varphi^*)}{3} \right) - Log\ \frac{|W(z)|^2}{4} \tag{2}$$

†Most of the material presented here has been worked out in collaboration with C. López.

In particular the scalar potential takes the form:

$$V(\varphi)= -e^{-G(\varphi,\varphi^*)}(3 + (G''_{\varphi\varphi^*})^{-1} |G'_\varphi|^2) + gauge\ terms \qquad (3)$$

The "minimal" kinetic term is obtained for $G''_{\varphi\varphi^*} = -1$, in which case one obtains

$$V(\varphi) = e^{\varphi\varphi^*/M^2}\left[|w_{,\varphi} + \frac{W\varphi^*}{M^2}|^2 - 3\frac{|W|^2}{M^2}\right] + gauge\ terms \qquad (4)$$

where $M^2 = m_p^2/8\pi$. Notice that in the limit $M \to \infty$ one recovers the global susy potential $= |W_{,\varphi}|^2$. Local supersymmetry is spontaneously broken if $\tilde{W}_{,\varphi} \equiv W_{,\varphi} + \frac{W\varphi^*}{M^2} \neq 0$ and $W(\varphi) \neq 0$ at the minimum of $V(\varphi)$. The goldstino is then swallowed by the gravitino, which gets a mass $m_{3/2} \sim e^{-G/2} \sim \frac{|W|}{M^2}$.

One may build specific SU(5) models[9-12] coupled to $N = 1$ supergravity by proposing different possible W's and chiral superfields content. No matter what model one has, formula (4) shows that all the scalars in the theory get (masses)^2 $\sim \frac{|W|^2}{M^4}$.[13,14] Since we do not want to have W-S doublets heavier than $\sim M_w \sim 100\,GeV$, we must impose $\frac{|W|}{M^2} \sim m_{3/2} \lesssim M_w$. This suggests setting $M_{3/2} \simeq M_w$ in which cases $\tilde{W}_{,\varphi} \sim M_w M$ and supersymmetry is broken at an intermediate scale $\sim 10^{10} GeV$, the "geometric scale". Thus large supersymmetry breaking[6-8] appears in a natural way in the context of local susy-GUT's.

In this scheme, the supersymmetry breaking takes place in a SU(3) x SU(2) x U(1) singlet "hidden sector" and the usual non-singlet particles feel the breaking only through explicit terms coming from the breaking of N = 1 supergravity. These terms which break susy explicitly but softly are the following:[9]

i. Masses for the scalars of order $M_\varphi \sim \frac{|W|}{M^2} \sim m_{3/2}$

ii. Trilinear scalar couplings coming from terms $\sim M_\varphi W$ with $M_Q \sim m_{3/2}$

iii. Gaugino masses $\sim m_{3/2}$.

The origin of the terms i) and ii) may be easily seen from Equation (4). In the case of minimal kinetic terms whose two terms (m_φ and M_φ) are related in each specific model. In a more general situation (non-minimal $G(\varphi\cdot\varphi^*)$) those two terms are independent.[15]

There are tree level gaugino masses if there is a non-minimal kinetic term for the Yang-Mills superfields in Equation (1).[13] If $f_{ab}(\varphi)$ is linear in the goldstino superfield, gauginos get masses $\sim m_{3/2}$.[9] If there are no tree level masses, radiative gaugino masses coming from heavy chiral superfields may be generated.

One may try to use these soft terms in order to break SU(2) x U(1) at a scale $\sim m_{3/2}$ at the tree level.[16] Since all the (masses)^2 of the scalars are positive one has to use large trilinear scalar couplings. However, this generally leads to undesired (charge breaking) minima.[17] Moreover, one generally light singlets which (if coupled to the W-S doublets) destabilize the gauge hierarchy.[18]

Thus the breaking of SU(2) x U(1) at the tree level through supergravity looks complicated.

An alternative to three level SU(2) x U(1) breaking was suggested in references 9-11. The key point is to realize that the soft terms breaking susy get renormalized when going from the grand unification scale to the weak scale, so that the effective scalar potential at low energies may be substantiality altered by radiative corrections.[11,12] The most relevant terms for the low energy scalar potential are the masses of the scalars. In particular, a negative (mass)^2 of the W-S doublets may signal the breaking of the SU(2) x U(1) symmetry. To see whether or not this happens, one has to consider the renormalization group equations for the explicit soft terms induced by supergravity. At the grand unification mass, (or, more precisely, at the Planck mass) all the scalars have equal masses $m^2_{\tilde{q}} = m^2_{\tilde{t}} = m^2_{H_1} = m^2_{H_2} \sim m^2_{3/2}$ and equal trilinear couplings. Gaugino masses are also equal and $\sim m_{3/2}$. The renormalization group equations with these boundary conditions were first considered in reference.[11] As an example, let us write down the evolution equation for the (mass)^2 of the right-handed scalar top, $\tilde{t}_R$:

$$\nu \frac{dm^2_{\tilde{t}_R}}{d\nu} = - \frac{\alpha_3}{(4\pi)} \frac{32}{3} M_3^2 - \frac{\alpha_1}{(4\pi)} \frac{32}{9} M_1^2 + \frac{h_t^2}{(4\pi)^2} 4 \left[m_t^2 + m^2_{\tilde{t}_R} + m^2_{\tilde{t}_L} + \mu_2^2 \right] \tag{5}$$

where M_3 and M_1 are the gluino and wino masses, M_t is the trilinear scalar coupling in the term

$$(h_t M_t) \tilde{t}_L \tilde{t}_R H_2 + h.c. \tag{6}$$

induced by supergravity and μ_2 is the Higgs doublet mass. For not too large h_t values the right-hand side of Equation 5 is negative so that the value of $m^2_{\tilde{t}_R}$ will increase as one goes to small energies. The same happens for the rest of the squarks. For the case of the sleptons the (mass)^2 remains essentially unchanged since the Yukawa couplings are small and there is no α_3 term. The evolution equation for the masses of the Higgs scalars is (neglecting small Yukawa couplings):[11]

$$\nu \frac{d\mu^2_{2,1}}{d\nu} = \frac{h^2_{t,b}}{(4\pi)^2} 6 \left[M^2_{t,b} + \mu^2_{2,1} + m^2{}_{\tilde{t}_L,\tilde{b}_L} + m^2{}_{\tilde{t}_R,\tilde{b}_R} \right] - \frac{\alpha_2}{4\pi} 6 M_2^2 - \frac{\alpha_1}{4\pi} 2 M_1^2 \tag{7}$$

Now the situation is completely reversed compared to that in Equation 5. The first four terms in the r.h.s. of Equation 7 may easily dominate the gauge interactions of the last two terms (at least in the case of μ_2). This is so for several reasons. First, h_t may be as large as g_2 and second, whereas M_2 and M_1

decrease at low energies, M_t^2 and m_t^2 do increase at small energies so that, as a whole, the first four terms in (7) will dominate for a heavy enough top quark. In this case the value of μ_2^2 will decrease at low energies and eventually will become negative leading to SU(2) x U(1) symmetry breaking.[11]

Indeed this actually happens for very reasonable values of the parameters involved (scalar masses, trilinear scalar couplings, gaugino masses). We have recently worked out a detailed numerical analysis of the relevant renormalization group equations for different boundary conditions and the results will be published elsewhere.[19] Let us present here some of the results. In Figure 1 we show the evolution of the scalar (mass)^2 for the third generation and the Higgses (similar results are obtained for the other generations) for the initial values $m = M_t = M_{\tilde{g}} = 100\ GeV$. In order to have the breaking at the appropriate scale one must impose the conditions:

$$M_w(M_w^2) \simeq 78\ GeV$$

The effective potential of the Higgses is of the form:

$$V(H_1, H_2) = \frac{1}{2}\left(D_2^2 + D_y^2\right) + \mu_1^2|H_1|^2 + \mu_2^2|H_2|^2 + \mu_3^2(H_1H_2 + h.c.) \tag{8}$$

The last term is required in order that H_1 acquires a v.e.v. and also to avoid the existence of an axion. It may come from the specific SU(5) superpotential which splits the SU(3) triplets and the SU(2) doublets in the 5 + $\bar{5}$ of SU(5). One could have additional quartics coming from a light singlet coupled to the doublets but, as explained above, light singlets would be dangerous. If μ_3^2 is not too large compared to $\left(\mu_1^2 + \mu_2^2\right)$ one obtains:

$$M_w^2 \simeq 2\cos^2\vartheta_w|\mu_2|$$

hence one must impose

$$\mu_2^2\left(Q^2 = M_w^2\right) \simeq -(62\ GeV)^2 \tag{9}$$

In the case shown in Figure 1, one obtains SU(2) x U(1) breaking at the appropriate scale for $h_t(M_x) = 0.2$ which corresponds to a top quark mass $m_t \simeq 100\ GeV$. One may need different values for the top quark mass depending on the values of the rest of the parameters involved. One gets smaller values for m_t if the gauginos are heavier. Also smaller values for m_t are obtained if the scalar masses at M_x (related in the minimal case to the gravitino mass) get smaller (this applies for not too large trilinear scalar couplings). To have an idea of the typical values of the parameters involved, we plot in Figure 2, the gaugino masses (at M_x) versus the values of m_t required to break SU(2) x U(1) at $\sim 80\ GeV$ for different values of the scalar masses at M_x. It does not seem possible to break SU(2) x U(1) for $m_t \lesssim 55\ GeV$ and to reach values for $m_t \sim 60-80\ GeV$ requires large values for gaugino masses (e.g. renormalized gluino masses of order 300-700 GeV). The results shown in Figure 2 correspond to the case $M_t = m$, equal scalar masses and trilinear couplings at Mx. For larger trilinear scalar couplings one

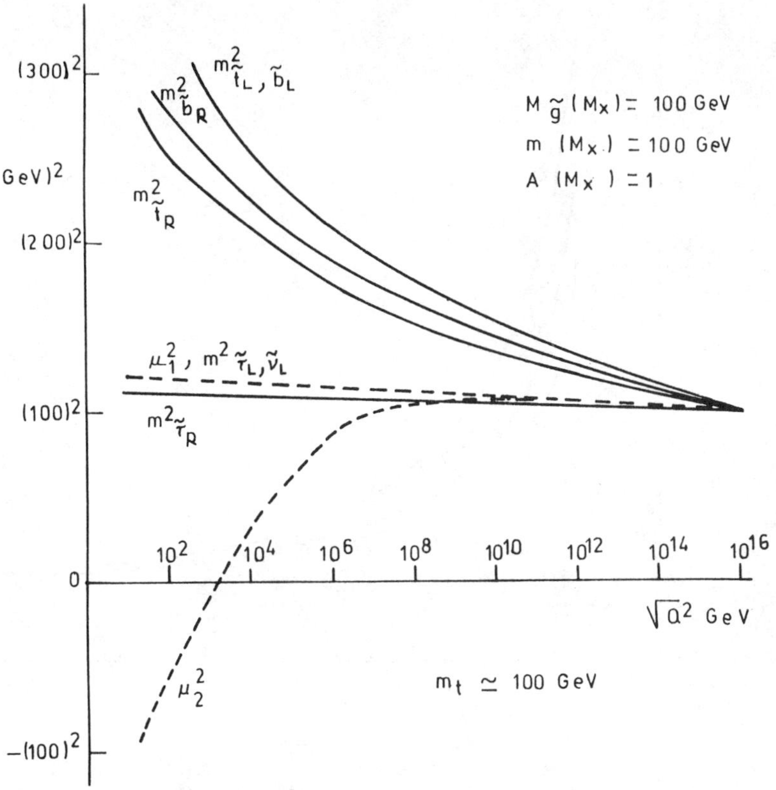

Figure 1. Evolution of the $(\mathrm{mass})^2$ of scalars as a function of the scale. The results correspond to the boundary conditions at $M_x : m = M_t = M_{\tilde{g}} = 100 GeV$. $Su(2)\times U(1)$ is broken at the right energy for a top quark mass $m_t \simeq 100 GeV$.

can manage to break SU(2) x U(1) with smaller gaugino masses. In any case the lower bound for the top quark mass applies:

$$m_t \gtrsim 55\ GeV$$

If renormalized gluino masses are lighter than $\sim 100\ GeV$, one needs $m_t \gtrsim 100\ GeV$ (in general even larger values) in order to get SU(2) x U(1) breaking at the appropriate scale.

The supersymmetric particle spectrum in this supergravity scheme is very similar to the one in radiative global susy-GUT's.[6-8] Since all the squarks and sleptons of different families will be (approximately) degenerate there are negligible flavor changing neutral currents. The only novel feature at low energies is the existence of the trilinear scalar couplings associated to each Yukawa coupling. These may have interesting phenomenological consequences.

Let us finally remark that this scheme for SU(2) x U(1) breaking is indeed very similar to the one proposed in Reference 3 in the context of global susy-

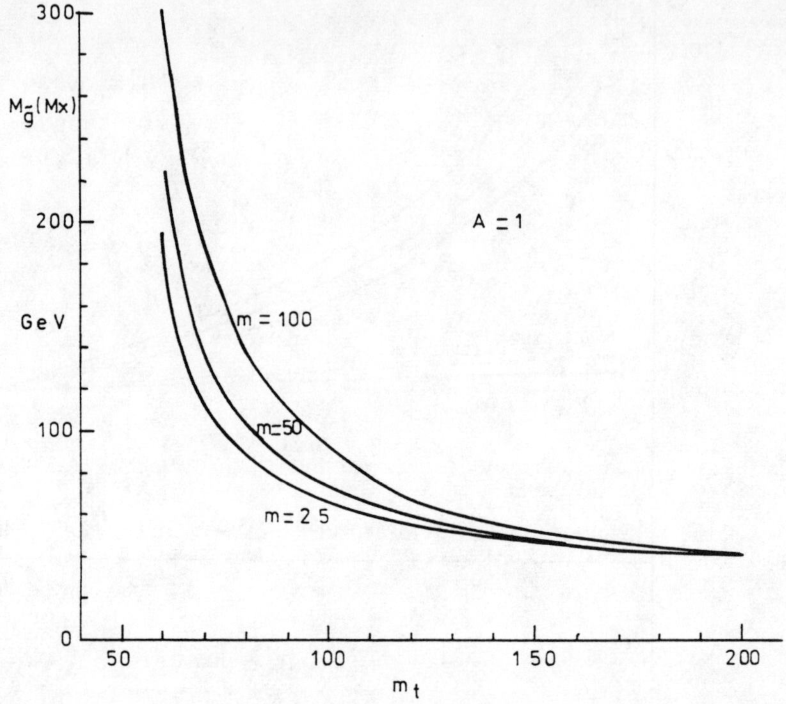

Figure 2. Gaugino masses at Mx as a function of the top quark mass for different values of the mass of the scalars at Mx. The minimum possible value for the top quark mass is $m_t \geq 55\,GeV$. The results shown correpond to $m = M_t$ at Mx (equal scalar masses and trilinear couplings).

GUT's. Also, this type of mechanism and, in particular, the renormalization group equations have been considered in the context of explicitly broken global susy by the authors of Reference 20. As mentioned above we have performed a thorough numerical analysis of the evolution equations for the soft terms induced by supergravity, which will be published elsewhere.[19] We understand that similar calculations have been performed by Alvarez-Gaumé, Polchinski and Wise[21] and by Ellis, Hagelin, Nanopoulos and Tamvakis.[22]

ACKNOWLEDGEMENTS

I would like to thank Professor G. Farrar and F. Henyey for their generous hospitality.

REFERENCES

1. Y.A. Gol'fand and E.P. Likhtman, Pis'ma Zh. Eksp. Teor. Fiz. **13**, 323 (1971). D. Volkov and V.P. Akulov. Phys. Lett. **46B**, 109 (1973). J. Wess and B. Zumino. Nucl. Phys. **B70**, 39 (1974).
2. S. Dimopoulos, S. Raby and F. Wilczek. Phys. Rev. **D24**, 1681 (1981). E. Witten. Nucl. Phys. **B186**, 513 (1981). S. Dimopoulos and H. Georgi. Nucl. Phys. **B193**, 150 (1981). N. Sakai. Z. fü Phys. **C11**, 153 (1982). L.E. Ibánez and G.C. Ross. Phys. Lett. **105B**, 439 (1981). H.P. Nilles and S. Raby. Nucl. Phys. **B198**, 102 (1982). M.B. Einhorn and D.R.T. Jones. Nucl. Phys. **B196**, 475 (1982).
3. L.E. Ibánez and G.C. Ross. Phys. Lett. **110B**, 227 (1982).
4. L. Alvarez-Gaumé, M. Claudsond and M. Wise. Nucl. Phys. **B207**, 96 (1982).
5. M. Dine and W. Fischler. Phys. Lett. **110B**, 227 (1982). C.R. Nappi and B.A. Ovrut. Phys. Lett. **113B**, 175 (1982).
6. J. Ellis. L.E. Ibánez and G.C. Ross. Phys. Lett. **113B**, 283 (1982).
7. R. Barbieri, S. Ferrara and D.V. Nanopoulos. Z. für Phys. **C13**, 267 (1982). M. Dine and W. Fischler. Princeton preprint (1982). S. Dimopoulos and S. Raby. Los Alamos preprint LA-UR-82-1282 (1982). J. Polchinski and L. Susskind, SLAC preprint SLAC-PUB - 2924 (1982).
8. J. Ellis, L. Ibánez and G.G. Ross. CERN preprint TH-3382 (1982).
9. L.E. Ibánez. Phys. Lett. **118B**, 73 (1982).
10. R. Arnowitt. P. Nath and A.H. Chamseddine. Phys. Rev. Lett. **49**, 970 (1982); and Northeastern preprint NUB 2565 (1982). R. Barbieri, S. Ferrara and C.A. Savoy. CERN preprint TH3365 (1982). B. Ovrut and J. Wess. Phys. Lett. **112B**, 347 (1982).
11. L.E. Ibánez, Universidad Autónoma de Madrid preprint FTUAM/82-8.
12. J. Ellis, D. Nanopoulos and K. Tamvakis. CERN preprint TH-3418 (1982). H.P. Nilles, CERN preprint TH-3398.
13. E. Cremmer, B. Julia, J. Scherk, S. Ferrara, L. Girardello and P. van Nieuwenhuizen. Nucl. Phys. **B147**, 105 (1979). E. Cremmer, S. Ferrara, L. Girardello and A. van Proeyen, CERN reprint TH-3348 (1982). J. Bagger and E. Witten, Princeton University preprint (1982).
14. J. Ellis and D. Nanopoulos, CERN Preprint TH-3319 (1982).
15. L. Hall, J. Lykken and S. Weinberg, Texas preprint.
16. R. Barbieri et al. in ref. 10; H.P. Nilles, M. Srednicki and D. Wyler, CERN preprint TH-3432 (1982).
17. J.M. Frere, D.R.T. Jones and S. Raby. Michigan preprint UM HE 82-58.
18. H.P. Nilles, M. Srednicki and D. Wyler, CERN preprint TH-3461.
19. L.E. Ibánez and C. López, Universidad Autónoma de Madrid preprint FTUAM/83-2.
20. K. Inoue, A. Kakuto, H. Komatsu, S. Takeshita. Kyushu University preprint, KYUSU-82-HE-5 (1982).
21. L. Alvarez-Gaumé, J. Polchinski and M. Wise. Harvard preprint.
22. J. Ellis, J. Hagelin, D. Nanopoulos and K. Tamvakis; see the talk by J. Ellis in these proceedings.

LIGHT-CONE SUPERSPACE AND THE FINITENESS OF THE N = 4 MODEL

Stanley Mandelstam
University of California, Berkeley, CA 94720

ABSTRACT

Superspace in light-cone co-ordinates takes a simple form. No auxiliary fields are necessary, and application to extended supersymmetries is straightforward. In a certain form of the light-cone gauge , the perturbation expansion of the N = 4 model is free of ultraviolet divergences. As a consequence, the β-function vanishes in any order of perturbation theory in any gauge.

INTRODUCTION

I should like to discuss superspace in light-cone co-ordinates . The Wess-Zumino algebra has been quoted in a paper by Siegel and Gates.[1] It is simpler than in a relatively covariant approach. This advantage is probably outweighed by the disadvantages of lack of manifest Lorentz covariance, and a relativistically covariant superspace is preferable for the models where it exists. Such covariant superspaces have not been found for extended supersymmetries, and arguments have been given that they do not exist. With light-cone co-ordinates, on the other hand, there is no difficulty in constructing extended superspaces.

In addition to the simplest supersymmetric model, there exist models where N, the number of Majorana fermions, is equal to 2 or 4. The N = 4 model, originally proposed by Gliozzi, Olive and Scherk, [2] was conjectured by Gell-Mann and Schwarz [3] to be finite in any order of perturbation theory. This conjecture has been verified up to three loops by Grisaru, Roček and Siegel and by Tarasov.[4] Related work has been performed by Ferrara and Zumino, by Sohnius and West, and by Stelle.[5] Once one has a superspace formulation of a supersymmetric theory, it often hapens that cancellations which previously appeared "miraculous" occur in a natural way. The vertex renormalization in the Wess-Zumino model is a well-known example. The wave-function renormalization, and therefore the coupling -constant renormalization, is infinite in the Wess-Zumino model. The N = 4 model is a gauge theory and, since we are using a physical gauge, we have the "naive" Ward identity which relates the vertex and wave-function renormalizations. We shall thereby be able to prove that all renormalization constants are perturbatively finite in a special form of the light-cone gauge.

The difficulty in obtaining covariant extended superspace is connected with the lack of correspondence between the number of fields and the number of particles. In any supersymmetric model, the particle multiplets will form a representation of the supersymmetry algebra. In particular, the number of bosons and fermions will be equal. When going from fields to particles this correspondence is lost, and one restores the balance by adding auxiliary fields. In the N = 4 model, there is one massless vector multiplet, six scalar multiplets and four fermion multiplets. Thus, if we count each helicity state separately, there are eight boson multiplets and eight fermion multiplets. Passing from particles to fields, we add two polarization states for the bosons and double the number of fermions. We thus have ten boson multiplets and sixteen fermion multiplets. No set of auxiliary fields has been found to restore the balance. With light-cone co-ordinates , the number of fields and particles is the same, and no problem arises. In fact, Green and Schwarz [6] encountered no problem in obtaining a manifestly supersymmetric formulation of the ten-dimensional string; the reason was that they were using light-cone co-ordinates.

WESS-ZUMINO MODEL

We shall illustrate the method by first discussing the Wess-Zumino model. In light-cone co-ordinates, [7] the Dirac equation for the lower two components of the four-spinor does not involve the "time" co-ordinate $x^+ = \frac{1}{\sqrt{2}}(x^0 + x^3)$, but only the "space" co-ordinates $x^1, x^2, x^- = \frac{1}{\sqrt{2}}(x^0 - x^3)$. Thus the two lower co-ordinates can be eliminated in favour of the two upper co-ordinates. The kinetic Lagrangian for the two-spinors is then quadratic. The supersymmetry transformations similarly split into two groups. They are most easily expressed in terms of the fields $A = \frac{1}{\sqrt{2}}(A_1 - iA_2)$ where A_1 and A_2 are the scalar and pseudo-scalar fields, and the helicity eigenstates $\psi = \frac{1}{\sqrt{2}}(\psi_1 + i\psi_2)$ of the spinor fields. The first group of supersymmetry transformations is:

$$\delta A = i\alpha^* \psi, \tag{1a}$$

$$\delta\psi = 2p^+ i\alpha A, \tag{1b}$$

where α is an infinitesimal anti-commuting c-number. We use momentum space for the x^- co-ordinate throughout; thus

$$p^+ = i\frac{\partial}{\partial x^-}, \quad (p^+)^{-1} f(x^-) = -\frac{i}{2}\int dx'^- \epsilon(x'^- - x^-) f(x'^-). \tag{2}$$

The second group of supersymmetry transformations connects A with the lower components of ψ, and therefore implicitly with the upper components. Since the Lorentz transformation between two different light-cone frames connect upper and lower components, we may write the transformations in the second group as a commutator between the Lorentz generators and the transformations in the first group. A Lagrangian which is invariant under both Lorentz transformations and supersymmetry transformations in the first group is automatically invariant under supersymmetry transformations in the second group. It is therefore unnecessary to consider the second group explicitly.

We may write the generator of the general supersymmetry transformation (1) as

$$\sum_{i=1}^{2} \overline{\alpha}_i Q_i, \qquad \alpha = \alpha_1 + i\alpha_2. \tag{3}$$

The Q's then have the simple anti-commutation relations

$$\{Q_i, Q_j\} = 4p^{+}\delta_{ij}. \tag{4}$$

To construct a superspace, we note that the first-quantized model has operators x, p together with the two supersymmetry operators Q_1 and Q_2 which connect the bosons with the fermions. After second quantization, the wave-functions become fields which are functions of co-ordinates corresponding to x and the Q's. One requires only one co-ordinate corresponding to each pair of conjugate variables, so that we have four commuting c-numbers x^{μ} and one anti-commuting θ. Our approach thus differs from the usual approach, which would have introduced a θ and a $\overline{\theta}$, and would have used constrained fields. The fields we use are unconstrained. A treatment of light-cone superspace along more conventional lines has been given by Brink, Lindgren and Nilsson, who have recently shown that their formalism can be used, in conjunction with the modified light-cone gauge used here, to prove finiteness [8].

Corresponding to Q_1 and Q_2, we define the two operators:

$$Q_2: \quad D = i\{\frac{\partial}{\partial\theta} - 2p^{+}\theta\}, \tag{5a}$$

$$Q_1: \quad \tilde{D} = \frac{\partial}{\partial\theta} + 2p^{+}\theta. \tag{5b}$$

We notice that the operators do satisfy the commutation relations (4) and, in particular, that

$$D^2 = \tilde{D}^2 = 2p^{+}. \tag{5c}$$

The superfield is defined as follows:

$$\phi = -i(2p^+)^{-1}\psi + \theta A. \tag{6a}$$

Note that the superfield is fermionic. We could have defined a bosonic superfield, but the above choice leads to a closer correspondence between the formulas of the present model and those of the N = 4 model. It is easily checked that the operators (5), when applied to (6), do effect the transformation (1).

We can now define the conjugate field

$$\phi^\dagger = -i(2p^+)^{-1}\psi^\dagger + \theta A^\dagger. \tag{6b}$$

The operator D is real, but $\tilde{D}$ is imaginary. Hence, when applying the D's to $\phi^\dagger$, we must make the correspondence:

$$Q_2: \quad D\phi^\dagger, \qquad Q_1: -\tilde{D}\phi^\dagger. \tag{7}$$

In writing down a supersymmetric action, two factors must be close in mind. The first is that we cannot define covariant derivatives in our present formalism, since we have only a single θ at our disposal. The two operators D and $\tilde{D}$ anti-commute with one another, but they commute with themselves, of course. The second point to bear in mind is the reversal of sign that occurs when $\tilde{D}$ acts on the conjugate field $\phi^\dagger$. These two possible complications cancel one another if we make the following rule: *Any term in the Lagrangian can only contain factors with an even number of $\tilde{D}$'s acting on ϕ and an odd number on $\phi^\dagger$ or vice versa.* The changes induces in the Lagrangian by a supersymmetry transformation will then disappear on integration.

One might regard θ as a real variable. From the point of view of invariance under x-y rotations it is preferable not to do so, since, under such rotations, θ changes by a phase factor. According to (6), the superfields are analytic functions of θ, i.e., they are polynomials in θ but are independent of $\bar{\theta}$. The conjugation $\phi \to \phi^\dagger$ is defined as Hermitian conjugation of the fields but not of θ. All of our formulas will involve only θ, $\frac{\partial}{\partial\theta}$ or $\int d\theta$. It is in this sense that our superfields only involve a single θ, whereas the conventional formalism involves two variables, θ and $\bar{\theta}$, for each complex θ.

The Lagrangian is as follows:

$$\begin{aligned} \mathcal{L} =& (\partial^\mu \phi^\dagger)\tilde{D}(\partial_\mu \phi) - m^2 \phi^\dagger \tilde{D}\phi - \frac{\sqrt{2}i}{3} g\{\phi(\partial_l \phi)(2p^+ \phi)) - \phi^\dagger(\partial_r \phi^\dagger)(2p^+ \phi^\dagger))\} \\ & - \sqrt{2} g m \{(\tilde{D}\phi)(\tilde{D}\phi)\phi^\dagger + (\tilde{D}\phi^\dagger)(\tilde{D}\phi^\dagger)\phi\} + 2g^2(\tilde{D}\phi^\dagger)(\tilde{D}\phi^\dagger)\tilde{D}\{(\tilde{D}\phi)(\tilde{D}\phi)\}, \\ & [\partial_{r,l} = \partial_1 \pm i\partial_2]. \end{aligned} \tag{8}$$

This Lagrangian does satisfy the supersymmetry rule quoted above. In light-cone co-ordinates, Lagrangians always contain a quartic as well as a cubic term, since there are no auxiliary fields.

N = 4 MODEL

The N = 4 model has four times as many supersymmetry operators as the Wess-Zumino model; sixteen in the covariant formulation or eight in the light-cone formalism. We may divide them into four helicity-increasing elements $Q_{a\alpha}$ and four helicity-decreasing elements $Q_{b\alpha}$; α is a new four-valued index. Associated with, but not necessarily implied by, the supersymmetry invariance is a global SU(4) invariance between the indices α. In addition, the N = 4 model is a gauge theory with an arbitrary gauge group; the gauge symmetry is completely separate from the supersymmetry and the global SU(4) symmetry.

We may construct the supersymmetry multiplets by successive application of the helicity decreasing supersymmetry operators to the state of highest helicity, which is equal to 1. Since the supersymmetry operators anticommute, the states must be antisymmetric in the operators. We thus have

State	Helicity	
$\|1\rangle$	1	One vector multiplet
$Q_{b\alpha}\|1\rangle$	$\frac{1}{2}$	Four Majorana spinor multiplets
$Q_{b\alpha}Q_{b\beta}\|1\rangle$	0	Six scalar multiplets
$Q_{b\alpha}Q_{b\beta}Q_{b\gamma}\|1\rangle$	$-\frac{1}{2}$	
$Q_{b1}Q_{b2}Q_{b3}Q_{b4}\|1\rangle$	-1	

The states with helicity $-\frac{1}{2}$ and -1 complete the vector and Majorana spinor multiplets. The N = 4 model, unlike the Wess-Zumino model, is T.C.P. self-conjugate.

It is convenient to define the combinations

$$Q_{1\alpha} = Q_{a\alpha} + Q_{b\alpha}, \tag{9a}$$

$$Q_{2\alpha} = i\{Q_{a\alpha} - Q_{b\alpha}\}, \tag{9b}$$

which satisfy the anti-commutation relations

$$\{Q_{i\alpha}, Q_{j\alpha}\} = 4p^+\delta_{ij}. \tag{10}$$

We now represent the Q's in terms of four θ's as follows:

$$Q_{2\alpha}: \quad D_\alpha = i\left(\frac{\partial}{\partial\theta^\alpha} - 2p^+\theta^\alpha\right), \tag{11a}$$

$$Q_{1\alpha}: \quad \tilde{D}_\alpha = \frac{\partial}{\partial\theta^\alpha} + 2p^+\theta^\alpha. \tag{11b}$$

The D's satisfy the commutation relations

$$\{D_\alpha, D_\beta\} = \{\tilde{D}_\alpha, \tilde{D}_\beta\} = 4p^+\delta_{\alpha\beta}, \tag{11c}$$

$$\{D_\alpha, \tilde{D}_\beta\} = 0. \tag{11d}$$

The supersymmetry condition will then be that, for each α, any term in the Lagrangian can only contain factors with an even number of $\tilde{D}$'s acting on ϕ and an odd number on $\phi^\dagger$ or *vice versa*.

In writing down the superfield, we start from the highest helicity field and go downwards. Thus:

$$\begin{aligned}\phi =& i(2p^+)^{-1}V + (2p^+)^{-1}\theta^\alpha\psi_\alpha + \frac{i}{4}\theta^\alpha\theta^\beta\rho^\mu_{\alpha\beta}A_\mu \\ &+ \frac{1}{3!}\epsilon_{\alpha\beta\gamma\delta}\theta^\alpha\theta^\beta\theta^\gamma\psi^\delta + 2ip^+\theta_1\theta_2\theta_3\theta_4 V^\dagger.\end{aligned} \tag{12}$$

Each field carries a gauge index which has been suppressed. The positive- and negative-helicity vector fields and the six spinless fields have been represented by the symbols V, V† and A_μ. The matrix $\rho^\mu_{\alpha\beta}$ is the appropriate Clebsch-Gordan matrix for SU(4) [or SO(6)]; an explicit form has been given by Brink, Scherk and Schwarz [9]. We specify further that the ρ's satisfy the condition

$$\rho_{\alpha\beta} = -\epsilon^{\alpha\beta\gamma\delta}\rho^*_{\gamma\delta}. \tag{13}$$

Since the supermultiplet is self-conjugate, the fields ϕ and $\phi^\dagger$ are not independent. They are related by the equation

$$\phi^\dagger = (2p^+)^{-2}\tilde{D}\phi, \tag{14}$$

where

$$\tilde{D} = \tilde{D}_1\tilde{D}_2\tilde{D}_3\tilde{D}_4. \tag{15}$$

In proving (14) we must make the use of the condition (15).

The Lagrangian for the model is as follows:

$$\mathcal{L} =(\partial^{\mu}\phi).(\partial^{\mu}\phi) - \frac{\sqrt{2}ig}{3}\phi.\{(\partial_l\phi)\times(2p^{+}\phi)\} + \frac{\sqrt{2}ig}{3}\phi^{\dagger}.\{(\partial_r\phi^{\dagger})\times(2p^{+}\phi^{\dagger})\}$$

$$\frac{-g^2}{64}\sum_{\alpha}\{(\mathcal{D}_{\alpha}\phi)\times(\mathcal{D}_{\alpha}\phi)\}.(2p^{+})^{-2}\{(\mathcal{D}_{\alpha}\phi^{\dagger})\times(\mathcal{D}_{\alpha}\phi^{\dagger}).\}. \qquad (16)$$

All dot and cross products refer to the gauge degree of freedom. The field $\phi^{\dagger}$ is regarded as a function of ϕ according to (14).

As we have mentioned, the operator $\phi^{\dagger}$ in (16) is to be regarded as a function of ϕ defined by (15). On inserting (15) in (16), we find

$$\mathcal{L} = (\partial^{\mu}\phi)\cdot(\partial_{\mu}\phi) - \frac{\sqrt{2}ig}{3}\phi\cdot\{(\partial_l\phi)\times(2p^{+}\phi)\}$$

$$-\frac{\sqrt{2}ig}{3}(2p^{+})^{-2}\phi\cdot\prod_{\alpha}(2i\partial_{x^-,\theta^{\alpha}})\{[(2p^{+})^{-2}\partial_r\phi]\times(2p^{+})^{-1}\phi\}$$

$$-\frac{g^2}{64}\sum_{\alpha}(2i\partial_{x^-,\theta^{\alpha}})(\phi\times\phi)\cdot(2p^{+})^{-2}\prod_{\beta\neq\alpha}(2i\partial_{x^-,\theta^{\beta}})\{(2p^{+})^{-1}\phi\times(2p^{+})^{-1}\phi\},$$

$$(17)$$

where

$$2i\partial_{x^-,\theta}(\phi_1,\phi_2) = (2p^{+}\phi_1)\frac{\partial\phi_2}{\partial\theta} - \frac{\partial\phi_1}{\partial\theta}(2p^{+}\phi_2). \qquad (18)$$

It is understood that the four anti-commuting derivatives $\partial_{x^-,\theta^{\alpha}}$ and $\partial_{x^-,\theta^{\beta}}$ in the third and fourth terms (17) are to be written in cyclic order. The derivatives with respect to θ in (17) thus all occur combined with the symbol $\epsilon^{\alpha\beta\gamma\delta}$ and, since this is an SU(4) invariant combination, the Lagrangian is manifestly SU(4) invariant.

It is now a straightforward matter to obtain Feynman rules; as in the Wess-Zumino model, the fields are unconstrained.

ULTRA-VIOLET FINITENESS

To prove the vanishing of the β-function, we examine a general vertex diagram. Let us consider an external three-point vertex of the form of the second term of (17). By using the identity

$$\phi_A.(\partial_l\phi_B \times 2p^+\phi_C) + \phi_A.(\partial_l\phi_C \times 2p^+\phi_B)$$
$$= -(\partial_l\phi_A)\cdot(\phi_B \times 2p^+\phi_C) + (2p^+\phi_A)(\phi_B \times \partial_l\phi_C), \qquad (19)$$

we note that we can choose one pair of lines meeting at the vertex, including the external line, and apply the factors ∂_l and $2p^+$ to them. Thus the external line has at least one factor ∂_l and $2p^+$. We can treat the other terms of (11) similarly; *each external line has at least one factor of* ∂_1, ∂_r, $2p^+$ *or* $\frac{\partial}{\partial\theta}$.

We thus find that there are more powers of p on the external lines of the vertex corrections than on the external lines of the bare vertex. (A factor $\frac{\partial}{\partial\theta}$ is dimensionally equivalent to a factor $p^{\frac{1}{2}}$). It might therefore be expected that the number of powers of p on the internal lines is insufficient to give a divergence. The power counting is easly performed and confirms this result: *the vertex corrections are finite, provided it is legitimate to employ naive power counting, with all components of p treated equally.*

In the usual light-cone gauge, such power counting is *not* in fact permissible.[10] The reason is that ultra-violet divergences appear from the region where p^+ and $p^+p^- - \mathbf{p}^2$ are finite, while p^- and $\mathbf{p}^2$ are large. The poles in the factors $(p^+)^{-1}$ prevent us from continuing to imaginary p° and thus avoiding these dangerous regions.

The condition $A^+ = 0$ does not define the light-cone gauge uniquely, since it remains true under a gauge transformation which depends on x^i and x^+ but not on x^-. The ambiguity is reflected in the $i\epsilon$ prescription in the factors $(p^+)^{-1}$. Usually one takes a principal value prescription. If we could use the prescription $(p^+)^{-1} \to (p^+ + i\epsilon p^-)^{-1}$, there would be no difficulty in continuing the p° integration to imaginary p°. After such continuation, it is easy to see that naive power counting is valid.

It is not difficult to show that one can define a light-cone gauge with the above $i\epsilon$ prescription. Such a "modified light-cone gauge" is inconvenient for most purposes, since it is only invariant under Lorentz transformations which leave both p^+ and p^- unchanged. For our purposes this is the best gauge to use, since the vertex functions are finite. It is also easy to prove that all n-point functions with $n \geq 3$ are finite.

For the two-point function, the above reasoning would still allow a divergent term of the form $A\mathbf{p}^2\delta_{ij} + B\mathbf{p}_i\mathbf{p}_j$. To show that such a term does not in fact occur, we use the Ward identity in the form $\Lambda^i(p,p,0) = -\frac{\partial}{\partial p_i}\prod(p,p)$.

This version of the Ward identity is valid only if proper Green's functions involving gluons of all four polarizations are free of singularities when any of the p^+'s becomes zero; a condition which is true in the modified light-cone gauge (though not in the usual light-cone gauge). From the differential form of the Ward identity and the finiteness of Λ, we can conclude that divergent terms proportional to $\mathbf{p}^2\delta_{ij}$ or to $p_i p_j$ cannot occur in Π. The two-point function, and the complete model, are thus finite in any order of perturbation theory.

In other gauges the wave-function renormalization will generally not be finite. The divergence is a pure gauge artifact. The β-function will always vanish, however, since its vanishing is a gauge-invariant condition.

We may finally note that our method of constructing the superspace without $\bar{\theta}$'s can also be applied to the covariant approach. For a more detailed version of the material presented here, we refer the reader to Refs. 11 (Light-cone superspace and finiteness) and 12 (Covariant superspace).

ACKNOWLEDGMENT

Research supported by the National Science Foundation under grant number PHY-81-18547.

REFERENCES

1. W. Siegel and S.J. Gates, Nucl. Phys. **B189**, 295 (1981).

2. F. Gliozzi, D. Olive and J. Scherk, Nucl. Phys. **B122**, 253 (1977).

3. M. Gell-Mann and J. Schwarz, unpublished.

4. M. Grisaru, M. Roček and W. Siegel, Phys. Rev. Lett. **45**, 1063 (1980); A. Tarasov, (to be published).

5. S. Ferrara and B. Zumino (unpublished); M. Sohnius and P. West, Phys. Lett. **100B**, 245 (1981); K. Stelle, L.P.T.E.N.S. 81/24 (1981); Proceedings of the Paris Conference on High-Energy Physics (1982). The authors of the latter two references have informed me that they have succeeded in constructing a covariant N = 2 superspace, and thereby completing an alternative proof of the finiteness of the N = 4 model.

6. M. B. Green and J. H. Schwarz, Nucl. Phys. **B181**, 502 (1981).

7. J. Kogut and D. Soper, Phys. Rev. **D1**, 2901 (1970); J.D. Bjorken, J. Kogut and D. Soper, Phys. Rev. **D3**, 1382 (1971).

8. L. Brink, O. Lindgren and B.E.W. Nilsson, Göteborg preprint 82/21 and

preprint UTTG 1-82.

9. L. Brink, J.H. Schwarz and J. Scherk, Nucl. Phys. **B121**, 77 (1977).

10. J.M. Cornwall, Phys. Rev. **D10**, 500 (1974).

11. S. Mandelstam, Berkeley preprint 82/15; Nucl. Phys. (to be published).

12. S. Mandelstam, Berkeley preprint 82/13; Phys. Lett. (to be published).

LOCAL SUPERSYMMETRY AND THE PROBLEM OF THE MASS SCALES*

H. P. Nilles
Stanford Linear Accelerator Center
Stanford University, Stanford, California 94305

and

CERN
CH-1211 Geneva 23, Switzerland

ABSTRACT

Spontaneously broken supergravity might help us to understand the puzzle of the mass scales in grand unified models. We describe the general mechanism and point out the remaining problems. Some new results on local supercolor are presented.

To begin let us remind you of the specific motivation that led to the application of supersymmetry[1] to grand unified models. It was the problem of three mass scales that we did not understand in terms of each others. These were the Planck mass $M_P \sim 10^{19}$ GeV which we know from the gravitational interactions, the speculative grand unification scale $M_X \sim 10^{14}$-10^{16} GeV related to proton decay and $M_W \sim 100$ GeV, the breakdown scale of the weak interactions. A first inspection of these three scales shows that M_W is tiny compared to the other scales: $M_W \approx 0$. One would like to understand why this is the case. An explanation could be a symmetry that keeps M_W small. Since the breakdown of the weak interactions is realized through vacuum expectation values of fundamental scalar fields we have only one choice for such a symmetry: Supersymmetry. In order to have $M_W \neq 0$ supersymmetry should be (spontaneously) broken at a scale M_S that is related to M_W. Once such a relation is established there remains the question whether one has gained anything. One has replaced M_W by M_S and one has now to face the problem to explain M_S in terms of M_X and M_P. There have however been some improvements. The first is a technical improvement due to the special behavior of supersymmetric field theories. It gives us the possibility that mass relations established at the tree graph level remain stable in perturbation theory. Secondly it is possible that M_S is much larger than M_W. The ratio can be as big as $M_S/M_W \sim 10^9$ as shown in the most promising globally supersymmetric models.[2] Such a large scale, however, leads necessarily to the introduction of supergravity since the gravitino mass $m_{3/2}$ is given by M_S^2/M_P which is now as big as the

* Work supported by the Department of Energy under contract number DE-AC03-76SF00515.

weak scale. Spontaneously broken local supersymmetry has additional desired properties. It admits vanishing vacuum energy in the case of broken supersymmetry[3] (which was impossible in the global case). Through the superHiggs effect the goldstino is absorbed; and finally it admits the Planck scale M_P to appear explicitly in the model. Maybe the small scale M_W can only be understood in terms of the two scales M_X and M_P.[4]

One thus is tempted to apply $N = 1$ supergravity to the models of high energy physics.[5,6] $N = 1$ supergravity is a nonrenormalizable theory and we can at best use it as an effective Lagrangian approach with the central assumption that there exists a satisfactory theory of gravity (which we do not know yet) for which $N = 1$ supergravity is a (good) approximation. This final theory should provide us with a cutoff for our nonrenormalizable approximation.

In such a model we assume that the gravitino mass is of the order of the weak scale and there are now two questions to be answered:

(1) How does one obtain a small scale $m_{3/2}$?

(2) How does the breakdown of local supersymmetry induce the breakdown of the weak interactions ?

We restrict ourselves to models where $m_{3/2}$ is the only small scale and where the weak interactions are restored in the limit $m_{3/2} \to 0$. Before we start our discussion we give the necessary formulae following the work of Cremmer et al.[5] for a single chiral superfield (z,x,h) coupled to supergravity. The model is characterized by a Kähler potential ($M = M_P$)

$$G(z,z^*) = -\frac{k}{M^2} - \log\left(\frac{|g|^2}{M^6}\right) \tag{1}$$

where g(z) is the superpotential and

$$k = -3M^2 \log(-\phi/3) \tag{2}$$

represents the choice of kinetic terms. Globally "normal" kinetic terms correspond to[7]

$$\phi = -3\left(1 - \frac{zz^*}{3M^2}\right) . \tag{3}$$

To discuss the model in supergravity one often makes the simplifying assumption of minimal kinetic terms

$$\phi = -3 \exp\left(-\frac{zz^*}{3M^2}\right) \tag{4}$$

which we will also use throughout this paper. In the presence of supergravity the scalar kinetic terms become

$$M^2 \frac{\partial G}{\partial z \partial z^*} |\partial_\mu z|^2 = M^2 G_{1zz^*} |\partial_\mu z|^2 \quad . \tag{5}$$

The potential is given by

$$V = -M^4 \exp(-G)\left[3 + \frac{|G_{1z}|^2}{(G_{1zz^*})}\right] . \tag{6}$$

The transformation law for the chiral fermion is

$$\delta\chi = 2M^2 \exp(-G/2)\frac{G_{1z^*}}{\sqrt{-2G_{1zz^*}}}\varepsilon + \ldots \tag{7}$$

where the dots denote field dependent terms. Formula (7) is the relevant expression to decide whether supersymmetry is broken or not.

We can now compare the corresponding formulae in the global and local case (we use minimal kinetic terms). The potential

$$V_L = \exp\left(\frac{zz^*}{M^2}\right)\left[|F_L|^2 - \frac{3|g|^2}{M^2}\right] \quad ; \quad V_G = |F_G|^2 . \tag{8}$$

The order parameter

$$F_L = \frac{\partial g}{\partial z} + \frac{z^*}{M^2} g \quad ; \quad F_G = \frac{\partial g}{\partial z} . \tag{9}$$

The supersymmetry by breaking scale is given by

$$M_S^2 = |F_L| \exp\left(\frac{zz^*}{2M^2}\right) \quad ; \quad M_S^2 = |F_G| = E_{vac}^2 . \tag{10}$$

Observe that the relation $M_S = E_{vac}$ is no longer valid in the local case. The massless Goldstino is absorbed by the gravitino which has a mass

$$m_{3/2} = \frac{M_S^2}{\sqrt{3}\, M} .$$

This concludes our presentation of the notation.

The most exciting property of $N = 1$ supergravity is the possibility to induce large mass gaps. A large hierarchy of mass scales can appear in which two large masses induce a small mass. This behavior has first been observed in a pure Yang Mills gauge theory coupled to supergravity.[4,8] A condensation of the gauginos λ at a scale $\langle\lambda\lambda\rangle = \mu^3$ was shown to break local supersymmetry at a scale $M_S^2 \sim \mu^3/M$ resulting in a gravitino mass $m_{3/2} \sim \mu^3/M^2$. A scale μ as large as the grand unification scale M_X and the scale M induce a small gravitino mass of order of a few TeV. A closer look at the situation showed that in this case nothing else could have happened since we know that the $\lambda\lambda$-condensation does not break global supersymmetry.[9] The breakdown scale M_S^2 has to be suppressed by $1/M$ to disappear in the global limit $M \to \infty$.

Meanwhile the general coupling of Yang Mills interactions to N = 1 supergravity has been worked out.[6] Consider a pure gauge theory and a chiral superfield in supergravity.[10] The transformation of the chiral fermion now reads as follows

$$\delta\chi = \left(\exp\,(-G/2)\,G' + \frac{1}{4}\,f'(z)\,\lambda\lambda\right)\varepsilon \quad . \tag{11}$$

Suppose that G' vanishes at the minimum. A condensation of the gauge fermion breaks supersymmetry provided that $f'(z) = \partial f(z)/\partial z$ is nonzero. The function f(z) denotes nonminimal kinetic terms for the gauge interactions. Such nonminimal kinetic terms are known to exist in extended supergravities, whereas in N = 1 supergravity this function is a free parameter. Suppose we choose

$$f(z) = 1 + \sigma\,\frac{z}{M} \quad . \tag{12}$$

The condensate $\langle\lambda\lambda\rangle = \mu^3$ then leads to a breakdown of supersymmetry at the scale

$$M_S^2 = \frac{1}{4}\,\sigma\,\frac{\mu^3}{M} \tag{13}$$

and a gravitino mass

$$m_{3/2} = \frac{\sigma\mu^3}{4\sqrt{3}\,M^2} \quad . \tag{14}$$

This is true for general f as long as the vacuum expectation values of the scalar fields are not much larger than the Planck scale.

The condensate will in general lead to a cosmological constant as can be seen from the general form of the potential

$$V = -3\exp\,(-G) + \left|\exp\,(-G/2)\,G' + \frac{1}{4}\,f'\,\lambda\lambda\right|^2 \quad . \tag{15}$$

This cosmological constant can however be cancelled by the scalar sector at the cost of one fine tuning of parameters. This will be seen later in special examples.

The most important result is Eq. (14), $m_{3/2} \sim \sigma\mu^3/M^2$. In principle one could imagine such a relation to occur in models with only chiral superfields. Suppose one has a superpotential g(z) with an intrinsic scale μ. In general one would then expect to have at the minimum $g_o = \lambda\mu^3$ where λ is a coupling constant. If global supersymmetry would be broken this would lead to $M_S^2 \sim \lambda\mu^2$ not much different from the scale μ. Suppose now however that we have broken supergravity with vanishing vacuum energy. Formulae (8) and (10) then lead to

$$M_S = \frac{\sqrt{3}\,|g|}{M} \sim \frac{\lambda\mu^3}{M} \tag{16}$$

comparable to (13). Unfortunately nobody was able up to now to find a superpotential that leads to (16) where vanishing vacuum energy is achieved for the absolute minimum of the potential. The only known superpotential that has broken supersymmetry and $E_{vac} = 0$ at the absolute minimum is[5]

$$g(z) = m^2(z + \beta) \tag{17}$$

with $\beta = \pm(2 - \sqrt{3})M$. The supersymmetry breakdown is of order m and one has to choose $m \sim 10^{11}$ GeV to obtain a gravitino mass in the TeV range. In the dynamical case [Eq. (14)] the scale μ could, however, be as large as the grand unification scale to obtain $m_{3/2} \sim O(M_W)$.

Given now a scale $m_{3/2} \sim O(M_W)$ we are confronted with the question how to apply the supersymmetry breaking[11-20] to models of quarks, leptons and Higgs particles. The supersymmetry breakdown has to appear in general in a distant (hidden) sector in order to keep the splittings in the observable particle spectrum as small as $m_{3/2}$. This can be achieved of the hidden sector couples only gravitationally to the observable sector. The superpotential is split into two pieces, e.g.

$$g(z,L_i) = g(z) + g(L_i) \tag{18}$$

where z denotes the hidden fields and L_i the observable fields. We now want to discuss the question under which circumstances the breakdown of supergravity can induce the breakdown of the weak interactions and thus provide a link between $m_{3/2}$ and M_W. We will split this discussion into two parts and first discuss the $SU(3) \times SU(2) \times U(1)$ model in the TeV-region and later on include grand unified models.

In the first case the L_i in (18) denote light fields; quarks, leptons, etc. According to our motivation we demand the naturalness condition[15] that $g(L_i)$ be scale invariant. Superpotentials that violate this condition necessarily contain explicit small mass parameters (~100 GeV) for which we do not understand why they are small. This condition gives

$$\sum_i \frac{\partial g}{\partial L_i} L_i = 3g \quad . \tag{19}$$

The low energy theory contains besides the quark and lepton superfield Higgses in the $2 + \bar{2}$ representation of SU(2). This is however not enough if the naturalness condition is imposed since $g(L_i)$ would then consist only of the Yukawa couplings and have a Peccei Quinn symmetry under which H and $\bar{H}$ have the same charge. The simplest extension is the inclusion of a singlet Y with $YH\bar{H} + Y^3$ couplings.[15]

To discuss the low energy potential it is convenient to go to the flat limit $M \to \infty$, $m_{3/2}$ fixed.[13] One then arrives at a softly broken globally supersymmetric theory. The soft breaking term include gaugino masses, scalar masses and scalar trilinear couplings. If one restricts oneself to a pure scalar hidden sector with minimal kinetic terms and imposes (19) the low energy potential turns out to be[15]

$$V_{LE} = \left|\frac{\partial g}{\partial L_i}\right|^2 + Am_{3/2}(g+g^*) + m_{3/2}^2 |L_i|^2 \quad . \tag{20}$$

More general forms[13,20] can occur if one does not impose (19). A is a pure number and depends on the hidden sector. The sppearance of this terms breaks all R-symmetries that might have been present in the model. In the case under consideration the coefficient of the last term is universal. We will see later on that this need not be the case in general. Gaugino masses are generally expected[6] of order $m_{3/2}$, but could be smaller. Radiative corrections in connection with this masses break the universality of the $m_{3/2}^2|L_i|^2$ terms, such that even negative mass^2 for the Higgses might occur.[4,14] But let us first discuss the tree graph level situation. Since $g(L_i)$ is scale invariant we know that $V_{LE} = 0$ at the point where all fields have vanishing vacuum expectation values. To induce a breakdown of $SU(2) \times U(1)$ V_{LE} must admit solutions at negative energies. The trilinear term is the only term that can give negative contributions. For $A = 3$ one obtains from (20) using (19)

$$V_{LE} = |g_{1i} + m_{3/2} L_i^*|^2 \geq 0 \quad . \tag{21}$$

This shows that $A \geq 3$ is a necessary condition to induce an $SU(2) \times U(1)$ breakdown at the <u>tree graph level</u>, and this result is still independent of the special form of $g(L_i)$. A depends on the hidden sector. For $g = m^2(z+\beta)$ one obtains $A = 3 - \sqrt{3}$. Even of one allows more scalars in the hidden sector one usually obtains this value as long as one uses (18). A model with $A = 3$ has been found[16] by replacing (18) by

$$g(z,L_i) = m^2(z+\beta)\exp\left(\frac{g(L_i)}{m^2 M}\right) \quad . \tag{22}$$

This model has the nice property that A does not receive quadratic divergencies in one loop gravity corrections.[21] Generalizations of (22) have led to models with $A > 3$.

If one includes a strongly interacting Yang Mills theory in the hidden sector the situation becomes more general. The low energy potential reads[10]

$$V = |g_{1i}|^2 + Am_{3/2}(g+g^*) + Bm_{3/2}^2|L_i|^2 \tag{23}$$

with

$$A = -\sqrt{3}\,(z_o/M) \tag{24}$$

and

$$B = -2 + M^2|G_z'|^2 + \frac{1}{2}\left(G'\,\frac{f'\lambda\lambda}{4m_{3/2}} + \text{h.c.}\right) \quad . \tag{25}$$

The main difference is now that B can vary. Weak interaction breakdown at the tree level requires now either $A \geq 3\sqrt{B}$ or $B \leq 0$.

The cosmological constant induced by the condensate has to be cancelled by the scalar sector. Let us use the superpotential $g = m^2(z+\beta)$ to do this. We have explicitly computed two cases $\beta = 0$ and $\beta = 2M$. In both cases one is able to cancel the cosmological constant by adjusting m^2 to special values.[10] In the case $\beta = 0$ we obtain

$$A = -1.33 \quad ; \quad B = 1.59 \tag{26}$$

for $\beta = 2M$

$$A = \sqrt{3} \quad ; \quad B = -2 \tag{27}$$

It is evident that we can reach all values $1.59 \geq B \geq -2$ by varying β in the range $0 \leq \beta \leq 2M$. One thus can obtain models where B is close to zero. B remained still universal and this might cause problems as has been pointed out by Frère, Jones and Raby.[18] They showed that in models with $A \geq 3\sqrt{B}$ the absolute minimum usually corresponds to broken electric charge, due to the small Yukawa coupling that is responsible for the electron mass. This is a serious problem but it could perhaps be cured with cosmological arguments — the absolute minimum is separated by a high barier.[22]

This problem exists strictly because of the universality of B. This universality, however, is broken by radiative corrections which enhances the B values for quarks and leptons and reduces those for the Higgses of the model has a top quark mass larger than 20 GeV.[23] A model with A close to 3 and B_o close to one would not suffer from this problem. It might even be that B is changed to negative values by radiative corrections. Starting with $B = 1$ this however requires a large lower bound $m_t \sim 60$ GeV on the top quark mass.[24] It might therefore be useful to consider models with potential like (23) which have a small B.[10,19]

It is thus possible to construct models in which the gravitino mass induces the breakdown of the weak interactions, and which contain no small mass parameters. The breakdown of $SU(2) \times U(1)$ is solely induced by supergravity which can be read off from (23); in the limit $m_{3/2} \to 0$ $SU(2) \times U(1)$ is restored.

The next step is to include grand unification. The superpotential $g(L_i)$ now contains also heavy fields. For these heavy fields we allow explicit mass parameters μ of the order of the grand unification scale. For the light fields we however still impose condition (19). This discussion is still relevant here since there is a hidden way[12,20] to break this condition in grand unified models, which usually leads to a breakdown of $SU(2) \times U(1)$ through a fine tuning. Let us take a simple example to explain this

$$g = \lambda A_{24} H_5 \bar{H}_{\bar{5}} + m H_5 H_{\bar{5}} \quad . \tag{28}$$

In this model one has to fine tune to keep the Higgs doublets massless. One solves the equations $|g_{1i}| = 0$ and determines the vacuum expectation value of A_{24}. One then adjusts m in such a way that the Higgs doublets remain massless. This is the right way to fine tune in a globally supersymmetric model. In local supersymmetry one has to fine tune differently. The vev of A_{24} is determined from $|g_{1i} + (L_i^*/M^2)g| = 0$ and differs from the one in the global theory slightly by an amount of order $m_{3/2}$. If one now fine tunes in the global limit one induces a small $m\bar{H}H$ mass parameter in the local theory. This leads sometimes to a breakdown of $SU(2) \times U(1)$, but this is not a breakdown induced by supergravity. It can be removed by a slightly different choice of the fine tuning procedure. Models that avoid the fine tuning through group theoretical reasons[25] (like models with 75, 50 and $\overline{50}$ representations) immediately rule out this possibility. As a result, the low energy effective potential remains unchanged.

To discuss grand unified models we use a toy model with one light (L) and one heavy field (B) and a superpotential[26]

$$g = \mu B^2 + \lambda_1 B^3 + \lambda_2 B^2 L + \lambda_3 L^3 + \lambda_4 BL^2 \quad . \tag{29}$$

In the limit $M \to \infty$ and μ, $m_{3/2}$ fixed this leads to the following potential

$$\begin{aligned} V = & \left|2\mu B + 3\lambda_1 H^2 + 2\lambda_2 BL + \lambda_4 L^2\right|^2 \\ & + \left|\lambda_2 B^2 + 3\lambda_3 L^2 + 2\lambda_4 BL\right|^2 + Bm_{3/2}^2(|B|^2 + |L|^2) \\ & + Am_{3/2}(\lambda_1 B^3 + \lambda_2 B^2 L + \lambda_3 L^3 + \lambda_4 BL^2 + h.c.) \\ & + (A-1)m_{3/2}\ (\mu B^2 + h.c.) \end{aligned} \tag{30}$$

Observe that all splittings are of order $m_{3/2}$ except for the last term which is of order $\mu m_{3/2}$. Thus the heavy fields are split by a large amount, and one has to worry whether this might induce large splittings in the low energy sector through radiative corrections. The light fields are coupled to the heavy ones through the couplings λ_2 and λ_4. Let us discuss the two terms separately.

The effect of λ_4 can be seen in the tadpole of Fig. 1. It gives a contribution $\mu m_{3/2}(L^2 + L^{*2})$ in leading order. This is however cancelled by the graph in Fig. 2. What remains are contributions of order $m_{3/2}^2$. Graphs where a vertex $\mu\lambda_1$ (in Fig. 1) is replaced by $Am_{3/2}\lambda_1$ also give contributions of order $m_{3/2}^2(L^2 + L^{*2})$. Thus the hierarchy remains stable. The exercise shows however that $m_{3/2}$ cannot be much larger than the weak interaction scale M_W, even of the parameters A and B are very small.

We proceed to discuss the terms proportional to λ_2 (from $\lambda_2 B^2 L$). A graph like Fig. 3 which contains $(A-1)\mu m_{3/2}$ and $m_{3/2}\lambda_2$ terms explicitly is not cancelled by any other graph and contributes with $\mu m_{3/2}$ to the mass of the light particles. The reason is the light particle exchange in Fig. 3. In the previous case (λ_4) these contributions were suppressed by the heavy propagator. As a result the tree graph level hierarchy in our toy model is spoiled.

In realistic models this problem occurs only if the model contains light singlets that couple via $\lambda_2 LB^2$ to heavy particles, but most of the locally supersymmetric grand unified models have this disease. It appears in graphs like Fig. 4 where a light singlet connects the light Higgs doublets with the heavy triplets. The problem usually persists even in more complicated models that contain light singlets.[27] The only natural way to solve the problem with a light singlet is to impose a discrete symmetry that does not allow the feeddown of $\mu m_{3/2}$ to the light particles. Such a symmetry, however, makes it difficult to generate masses for the Higgs triplets, which however can be solved by imagination and a fairly large amount of group theory.[28] The cleanest way, however, is the absence of light singlets.

It seems at the moment that the marriage of localsupersymmetry and grand unification leads to phenomenologically acceptable models. It might even be possible to understand the small mass scale M_W in form of a conspiracy of the two big scales M_X and M_P as described. Ultimately one might understand M_X from gauged extended supergravities where the gauge interactions become strong at M_X and breaks the last remaining supersymmetry. The small gravitino mass can induce the breakdown of $SU(2) \times U(1)$ in a natural way. This could be achieved in several ways depending on the parameters of the special model. Those parameters, however, can only be fixed by new experimental input.

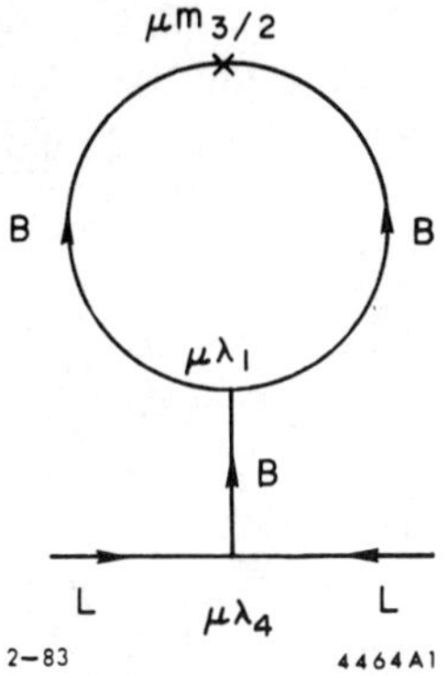

Fig. 1. Potentially dangerous contribution to light particle masses.

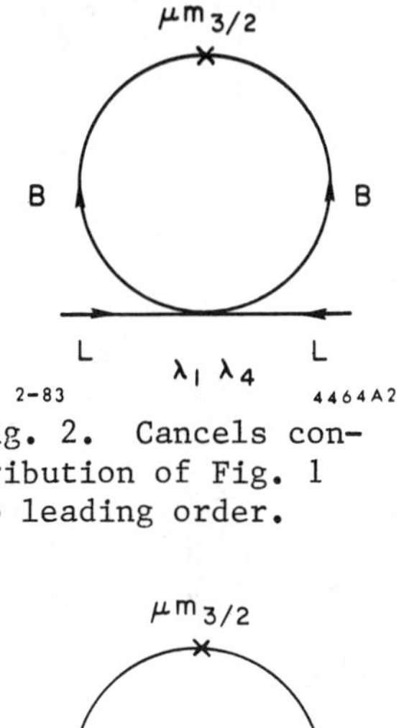

Fig. 2. Cancels contribution of Fig. 1 to leading order.

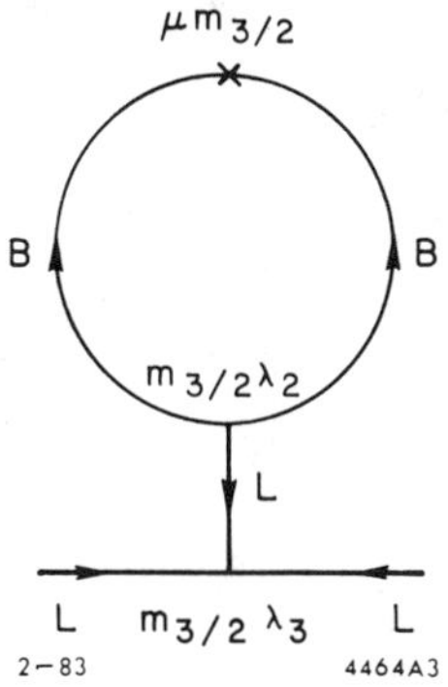

Fig. 3. Tadpole that spoils the hierarchy in the toy model.

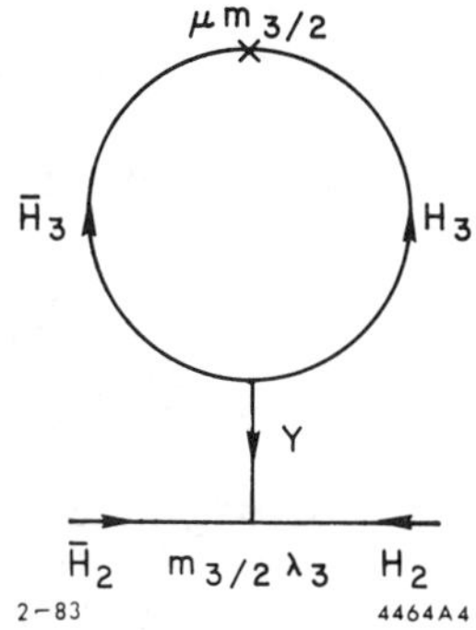

Fig. 4. The contribution that is relevant in grand unified models.

ACKNOWLEDGEMENTS

Various parts of the described work has been done in collaboration with S. Ferrara, L. Girardello, M. Srednicki and D. Wyler. I also would like to thank L. Hall, L. Ibanez, A. Masiero, S. Raby and S. Weinberg for discussions. Congratulations to G. Farrar and F. Henyey for the organization of this meeting.

REFERENCES

1. J. Wess and B. Zumino, Phys. Lett. 49B, 52 (1974).
2. R. Barbieri, S. Ferrara and D. Nanopoulos, Phys. Lett. 116B, 16 (1982); M. Dine and W. Fischler, Princeton IAS preprint (1982); S. Dimopoulos and S. Raby, Los Alamos preprint LA-UR-82 (1982); J. Polchinski and L. Susskind, SLAC preprint PUB-2924 (1982).
3. S. Deser and B. Zumino, Phys. Rev. Lett. 38, 1433 (1977).
4. H. P. Nilles, Phys. Lett. 115B, 193 (1982).
5. E. Cremmer, B. Julia, J. Scherk, S. Ferrara, L. Girardello and P. van Nieuwenhuizen, Nucl. Phys. B147, 105 (1979).
6. E. Cremmer, S. Ferrara, L. Girardello and A. Van Proeyen, Phys. Lett. 116B, 219 (1982) and CERN preprint TH-3348 (1982), to be published in Nucl. Phys. B.
7. H. P. Nilles, CERN preprint TH-3398 (1982), to be published in Nucl. Phys. B.
8. H. P. Nilles, Proceedings of the Johns Hopkins Workshop on Current Problems in Particle Theory 6, ed. G. Domokos et al., Florence (1982).
9. H. P. Nilles, Phys. Lett. 112B, 445 (1982).
10. S. Ferrara, L. Girardello and H. P. Nilles, CERN preprint TH-3494 (1982).
11. B. A. Ovrut and J. Wess, Phys. Lett. 112B, 347 (1982).
12. R. Arnowitt, A. H. Chamseddine and P. Nath, Phys. Rev. Lett. 49, 870 (1982).
13. R. Barbieri, S. Ferrara and C. A. Savoy, CERN preprint TH-3365 (1982), to be published in Phys. Lett. B.
14. L. Ibanez, CERN preprint TH-3374 (1982).
15. H. P. Nilles, M. Srednicki and D. Wyler, CERN preprint TH-3432 (1982), to be published in Phys. Lett. B120, 346 (1982).
16. E. Cremmer, P. Fayet and L. Girardello, LPTENS preprint 82/30 (1982).
17. J. Ellis, D. Nanopoulos and K. Tamvakis, CERN preprint TH-3418 (1982).
18. J. M. Frere, D. R. T. Jones and S. Raby, University of Michigan preprint HE 82-58 (1982).
19. M. Gaillard, L. Hall, B. Zumino, F. del Aguila, J. Polchinski and G. Ross, Berkeley preprint UCB-PTH-82/21 (1982).
20. L. Hall, J. Lykken and S. Weinberg, University of Texas preprint UTTG-1-83 (1983).
21. R. Barbieri and S. Cecotti, University of Pisa preprint SNS 9/1982.
22. L. Hall, private communication.

23. L. Ibanez and G. Ross, Phys. Lett. 110B, 215 (1982).
24. L. Ibanez and C. Lopez, L. Alvarez-Gaume, J. Polchinski and M. Wise, papers in preparation; see also the contribution of L. Ibanez to this conference.
25. J. Grinstein, Harvard preprint (1982); A. Masiero, D. Nanopoulos, K. Tamvakis and T. Yanagida, Phys. Lett. 115B, 375 (1982).
26. H. P. Nilles, M. Srednicki and D. Wyler, CERN preprint TH-3461 (1982); A. Lahanas, CERN preprint TH-3467 (1982).
27. S. Ferrara, D. Nanopoulos and C. Savoy, CERN preprint TH-3442 (1982).
28. A. Masiero and D. Wyler, private communication.

MAGNETIC MONOPOLES AND THE KALUZA-KLEIN THEORY

Malcolm J. Perry*
Dept. of Physics, Princeton University, Princeton, NJ 08544

ABSTRACT

It is shown that the Kaluza-Klein theory admits soliton solutions which are magnetic monopoles. These monopoles violate the principle of equivalence in the sense that their inertial mass is approximately the Planck mass, but their gravitational mass vanishes.

The Kaluza-Klein theory[1,2] was one of the first genuine attempts at the construction of a unified theory of electromagnetism and gravitation. In essence, it can be thought of a five-dimensional version of general relativity, where the spacetime M has metric $\gamma_{\mu\nu}$ ($\mu,\nu\ldots=1,2,3,4,5$), and signature (-++++). The action for this theory is the simplest possible generally covariant action namely

$$I= \int_m R(\gamma)\ (-\gamma)^{1/2}\ d^5x \tag{1}$$

(It should be noted that dimensionless units with $G=c=\hbar=1$ are used throughout this paper.) The obvious difficulty with this theory is that spacetime appears to be four-dimensional, not five-dimensional. However, if we suppose that one of the spatial directions is wrapped up in a very short distance scale, ℓ, then the world as we perceive it at energy scales very much smaller than ℓ^{-1} will be four-dimensional. It is commonly supposed that ℓ should be of the order of the Planck length.

The symmetries of the low-energy world in this rather simplified model are those of four-dimensional relativity and electromagnetism. Thus, the five-dimensional co-ordinate invariance, the five-dimensional diffeomorphism group has been broken by the compactification scheme down to the four-dimensional diffeomorphism group and the gauge group of electromagnetism.

$$\mathrm{Diff}^5 \xrightarrow[\text{Compactification}]{} \mathrm{Diff}^4 \times U(1). \tag{2}$$

The U(1) corresponds to a symmetry of the spacetime, which in this case is generated by translations along the compactified direction. Formally, there must abe a Killing vector $k^a \frac{\partial}{\partial x^a} = \frac{\partial}{\partial x^5}$ where $\frac{\partial}{\partial x^5}$ is a vector which is normal to the "four-dimensional spacetime." Under these circumstances we can pick a co-ordinate sysem which explicitly shows the degrees of freedom in this theory. The five-dimensional line element can be written as

$$ds^2 = V(dx^5 + \omega_i dx^i)^2 + \frac{1}{V}\, g_{ij}\ dx^i dx^j \tag{3}$$

($i,j\ldots=1,2,3,4$). V, ω_i and g_{ij} are the fundamental degrees of freedom;

*Supported by National Science Foundation Grant PHY80-19754.
0094-243X/84/11600121-05 $3.00

namely a scalar, a four-vector, and a metric tensor. Since $\frac{\partial}{\partial x^5}$ is Killing, v, w_i and g_{ij} can be chosen to be independent of x^5. All of these considerations can be readily generalized to higher dimensional spacetimes that produce unified theories with more interesting low-energy gauge groups.[3,4] In fact, one would nowadays regard this as the real motivation for Kaluza-Klein theories since it seems natural to formulate supergravity theories in higher dimensions, and to formulate string theories in ten dimensions. One could even argue that since string theory is the only theory of gravitation which is not known to be pathological, we should be working in ten dimensions ab initio.

In order to study our five-dimensional theory, we should start by searching for suitable classical solutins to the five-dimensional field equations. In the variables of equation (3), this amounts to extremizing

$$\int (V^{-1/2}R + V^{-3/2}\ \ V - \frac{5}{2} V^{-5/2} (\nabla V)^2 + \frac{1}{4} V^{5/2} F_{ij}F^{ij})(-g)^{1/2} d^4x. \quad (4)$$

where

$$F_{ij} = \nabla_i \omega_j - \nabla_j \omega_i \quad (5)$$

The covariant derivatives and the Ricci scalar R are formed from the four-dimensional metric g.

The simplest possibility, which corresponds to the vacuum, is to pick flat space, with V=1, $\omega_i=0$, $g_{ij}=\eta_{ij}$ (the metric four-dimensional Minkowski space), which has the topology of $R^4 x S^1$.

However, S^1 bundles over S^2 are classified by the integers, and so there exist many topologically inequiavalent possible metrics, corresponding the number of twists the fibration has. The first non-trival possibility is given by a single twist. The classical field equations are then solved by

$$\begin{aligned} g_{ab}dx^a dx^b &= dr^2 + r^2 d\theta^2 + r^2 \sin^2\theta s\phi^2 - Vdt^2 \\ w_a d_x{}^a &= 4m \cos\theta d\phi \\ V^{-1} &= 1+\frac{4m}{r} \end{aligned} \quad (6)$$

r, θ, φ are a set of polar co-ordinates, t is a time co-ordinate. m is a mass parameter. As $r \to \infty$, the metric becomes asympfatically flat. However, on the face of it, this five-dimensional metric is singular for two reasons. The first is that there is a string singularity along θ=π. Since φ is an azimuthal angle which must be identified with period 2π, it follows that this string singularity can be eliminated completely, provided that x^5 is also a periodic variable (which we assumed it was), but with the specific period of 16πm. Furthermore, at r=0, the scalar field V blows up. This is not a geometrical singularity since scalar polynomials of the curvature do not blow up there. The apparent singularity in $\gamma_{\mu\nu}dx^\mu dx^\nu$ at r=0 results merely from a poor choice of co-ordinates near the origin. One can change the co-ordinates, and find a locally flat set of co-ordinates at r=0 which is sufficient to demonstrate that r=0 is non-singular. (It should perhaps be noted that this metric is in fact Euclidean Taub-NUT solution[5] with an extra flat dimension added on in the time direction).

The electromagnetic field does not vanish for this solution. The electric field is

$$E_i dx^i = F_{oi}\, dx^i = 0 \tag{7}$$

and the magnetic field,

$$B_i dx^i = \frac{1}{2}\, \varepsilon_{ijk}\, F^{jk} dx^i = \frac{4m}{r^2}\, dr \tag{8}$$

Thus, the origin r=0 acts as a spherically symmetric point source of magnetic flux of strength P=4m. We have thus found a magnetic monopole which is a soliton in the Kaluza-Klein theory.

It should perhaps be noted parenthetically that Einstein and Pauli[6] thought that they had proved that solitons do not exist in the Kaluza-Klein theory. It turns out that this example evades the conditions of their theorem.

The inertial mass of such a soliton is given by the Arnowitt-Deser-Misner integral[7], which is determined by the asymptotic behavior of the spatial components of the metric. For this soliton, the inertial mass is m. Another measure of the mass of a soliton is the gravitational mass. This is best determined by looking at the behavior of slowly moving test particles near $r\to\infty$[8]. The motion of such particles is determined by the Newtonian gravitational potential. To calculate the motion of such particles, we note that test particles will follow geodesics in the five-dimensional metric $\gamma_{\mu\nu}$[9]. It turns out that the velocity in the x^5 direction determines the electric charge of these test particles. If we look at neutral objects, we discover tht the Newtonian potential is given by the r^{-1} part of γ_{oo}. However $\gamma_{oo}=1$ in our solution, so that the gravitational mass is zero. This is not as pathological as it seems: the principle of equivalence is violated. In theories with long-range scalar fields this is a common feature.[10]

As we have pointed out, momentum in the x^5-direction corresponds to electric charge. If we look at solutions of the Klein-Gordon equation, they will in general look like

$$\phi \sim e^{iqx^5} \times f(x^i) \tag{9}$$

where q is the electric charge of this mode. However x^5 is identified with period $16\pi m$, and ϕ must be single valued. Thus

$$q = \frac{n}{8m} \tag{10}$$

where n is an integer. Electric charge is quantized in units of e, thus we can derive an expression for the mass in terms of e

$$m = \frac{1}{4e} \tag{11}$$

In terms of traditional particle physics units, this is a mass $m = 1.27 \times 10^{20}$ GeV.

It is also possible to find solutions which contain N magnetic monopoles. In this case, the boundary at i^0 is an S^1 bundle over S^2 with N twists in it. In this case the metric is given by

$$g_{ij}dx^i dx^j = (dx^2+dy^2+dz^2) - Vdt^2 \tag{12}$$

$$V^{-1} = 1 + 4m \sum_{j=1}^{N} \frac{1}{|\vec{x} - \vec{x}_j|} \tag{13}$$

and

$$\text{grad } V^{-1} = \text{curl } \omega \tag{14}$$

x,y,z are co-ordinates in a flat Euclidean space, t is the Minkowskian time co-ordinate. The $j^{\underline{th}}$ monopole is located at $\vec{x}_j$, and $|\vec{x}-\vec{x}_j|$ is the Euclidean distance from $\vec{x}$ to $\vec{x}_j$. Grad and curl are the standard three-dimensional vector operations in flat space. It should be noted that these metrics are again completely non-singular, and that the mass of each monopole must be the same. Since $\vec{x}_j$ is arbitrary, the monopoles can be located anywhere in space. This means that there is no net force between the monopoles[5].

Finally, one can find solutions representing N antimonopoles by replacing (14) by

$$\text{grad } V^{-1} = -\text{curl } \omega \tag{15}$$

Further details can be found in the forthcoming paper by Gross and Perry[11]

REFERENCES

1. T. Kaluza, Sitz, Preuss Akad Wiss (Berlin), Math-Physik, Kl, 966, 1921.

2. O. Klein, Z. Physik 37, 895, 1926.

3. B.S. deWitt, "The Dynamical Theory of Groups and Fields", in "Relativity, Groups, and Topology" ets B.S. deWitt and C. deWitt, Gordon and Breach, New York, 1964.

4. E. Witten, Nucl. Phys. B186, 412, 1981.

5. S.W. Hawking, Phys. Lett. 60A, 81, 1977.

6. A. Einstein and W. Pauli, Ann. Math. 44, 131, 1943.

7. R. Arnowitt, S. Deser and C. Misner, Phys. Rev. 118, 1100, 1960.

8. S. Weinberg, "Gravitation and Cosmology" Wiley, New York, 1972.

9. A. Lichnerowicz, Théories Relativistes de la Gravitation et de l' Electromagnétisme", A. Masson et Cie, Paris 1955.

10. C. Brans and R.H. Dicke, Phys. Rev. 124, 925, 1964.

11. D.J. Gross and M.J. Perry, in preparation.

AN EFFECTIVE LAGRANGIAN FOR SUPERSYMMETRIC QCD*

Michael E. Peskin
Stanford Linear Accelerator Center
Stanford University, Stanford, California 94305

ABSTRACT

I present a Lagrangian which describes the spontaneous breaking of chiral symmetries in strongly interacting supersymmetric Yang-Mills theory with matter fields. This Lagrangian predicts that supersymmetry is spontaneously broken if the matter fields have precisely zero mass.

Over the past few years, as our understanding of weakly coupled supersymmetric theories has steadily increased, the dynamics of strongly coupled supersymmetric Yang-Mills theory has come to appear more and more mysterious. Initially, it was tempting to regard these theories as having qualitatively the same behavior as ordinary gauge theories of fermions. Using this hypothesis, Dine, Fischler, and Srednicki[1] and Dimopoulos and Raby[2] argued that these theories should show spontaneous supersymmetry breaking. This conclusion, however, was apparently contradicted when Witten derived a striking constraint on dynamical supersymmetry breaking.[3] This contradiction has left workers in this field more than a little puzzled and has led to a consensus that the pattern of chiral symmetry breaking in supersymmetric Yang-Mills theory must be an unusual one. However, it need not be so. In this lecture, I will demonstrate this by exhibiting an effective Lagrangian describing the spontaneous breaking of chiral symmetry in supersymmetric Yang-Mills theory which is consistent both with the physical picture of Dine, Fischler, Srednicki, Dimopoulos and Raby and with the constraints proved by Witten. This conclusion differs from that of a recent paper by Taylor, Veneziano, and Yankielowicz;[4] I will clarify the difference between my analysis and theirs as I proceed.

I will restrict my attention in this lecture to theories in which a gauge supermultiplet (A_μ, λ, D) couples to matter fields which belong to a real representation of the gauge group. For most of the analysis of this paper, I will take this representation to comprise n copies each of a complex representation r and its complex conjugate $\bar{r}$. The matter supermultiplets are, then, of the form:

$$(A_{ri}, \psi_{ri}, F_{ri}) + (A_{\bar{r}i}, \psi_{\bar{r}i}, F_{\bar{r}i})$$

where $i = 1, \ldots, n$. ψ denotes a left-handed fermion; the other fields are complex bosons. These models are essentially supersymmetric versions of QCD with n flavors; I will refer to them as SSQCD. They are the models to which Witten's theorem applies most directly.

---

*Work supported by the Department of Energy under contract DE-AC03-76SF-00515.

At the classical level, for zero mass matter fields, SSQCD has the global symmetry $U(n) \times U(n) \times U(1)$, where the last $U(1)$ corresponds to R-invariance. In the quantum theory, one $U(1)$ symmetry is destroyed by anomalies; the full global symmetry is, then, $U(n) \times U(n)$. One can give mass to the matter fields by adding to the Lagrangian a superpotential of the form:

$$W(A) = \sum_{i=1}^{n} m \, A_{ri} A_{\bar{r}i} \tag{2}$$

This potential breaks $U(n) \times U(n)$ explicitly to (vectorial) $U(n)$. In ordinary QCD, the formation of fermion pair condensates causes a spontaneous breaking of the chiral symmetry of the zero-mass theory; I see no good reason why this same physics should not appear also in the supersymmetric theory. Such fermion-pair condensates would give rise in SSQCD to the pattern of spontaneous symmetry-breaking:

$$U(n) \times U(n) \to U(n) \,. \tag{3}$$

I will argue that the symmetry-breaking pattern (3) is consistent with the constraints of supersymmetry by exhibiting a supersymmetric effective Lagrangian which gives a low-energy phenomenological description of this symmetry breaking. This Lagrangian should be the appropriate generalization to SSQCD of the description of the low-energy dynamics of QCD by a nonlinear sigma model.[5] More specifically, this Lagrangian has the following properties: First, it obeys a number of requirements which follow from exact properties of SSQCD:

1. The Lagrangian has the form

$$\mathcal{L} = \mathcal{L}_0 + tr\, m\Delta \tag{4}$$

   where $\mathcal{L}_0$ is invariant to $U(n) \times U(n)$ and Δ, which represents the matter-field mass term, transforms as an $(\bar{n}, n)$ under $U(n) \times U(n)$.
2. The Lagrangian is manifestly supersymmetric.
3. Supersymmetry is not spontaneously broken for any value of $m \neq 0$.

Requirement (3) follows from Witten's theorem.[3] Secondly, the Lagrangian is consistent with a number of intuitive requirements of the physical picture of chiral symmetry breaking by fermion pair condensates:

4. The pattern of spontaneous symmetry breaking is

$$U(n) \times U(n) \to U(n) \,;$$

   the associated Goldstone bosons appear as elementary fields of the phenomenological Lagrangian.
5. The gluino is heavy and irrelevant to low-energy physics; the gluino field does not appear in the Lagrangian.

6. The Lagrangian implies that $< \psi_r \psi_{\bar{r}} > \neq 0$ by insuring that, in the presence of the symmetry-breaking perturbation $tr\ m\Delta$, the Goldstone bosons receive $(mass)^2$ propertional to m.
7. The bosonic variables of the model live on a compact space.
8. The Lagrangian satisfies decoupling: sending one eigenvalues of m to infinity reduces the $U(n) \times U(n)$ version of the Lagrangian to the $U(n-1) \times U(n-1)$ version.

Since the requirements (6) and (7) are not completely obvious, and since they will play a crucial role in my analysis, I comment on them briefly. The authors of Refs. 1 and 2 argue that my assumption (6) already implies that supersymmetry is spontaneously broken. Their argument makes use of the Ward identity, valid if supersymmetry is manifest:

$$\left\langle \frac{1}{2}\psi_r \cdot \psi_{\bar{r}} + A_r \cdot F_{\bar{r}} \right\rangle = 0$$

If one eliminates F using the equations of motion, one finds

$$-\frac{1}{2} < \psi_r \cdot \psi_{\bar{r}} > = m < |A_r|^2 > \tag{5}$$

If $< |A|^2 >$ is regular as $m \to 0$, supersymmetry implies $< \psi_r \psi_{\bar{r}} > = 0$. But such regularity is not necessary, or even to be expected. In ordinary QCD one can cast $< \bar{\psi}\psi >$ into the form

$$< \bar{\psi}\psi > = m \left\langle tr \left(\frac{1}{-D^2 + m^2 + \frac{g}{4}\sigma_{\mu\nu}F^{\mu\nu}} \right) \right\rangle \tag{6}$$

where the expectation value is to be taken over configurations of the gauge field.[6] The object inside the trace is formally quite similar to the A_r propagator. If the right-hand side of (6) can remain nonzero as $m \to 0$, why should the right-hand side of (5) not also show this behavior?** I feel that the assumption (6) does not unduly prejudice the theory I will construct toward spontaneous supersymmetry breaking.

My assumption (7) would not be a strong assumption in ordinary field theories with global symmetries. However, in supersymmetric theories it is a very strong assumption, because supersymmetric nonliner sigma models with variables on compact spaces are not obtainable as limits of linear sigma models[7,8] I will simply assume that the nonperturbative dynamics of SSQCD gives rise to a compact manifold of possible vacuum states. It is here that my analysis differs from that of Taylor, Veneziano, and Yankielowicz[4]; those authors chose a set of dynamical variables which could be obtained from a supersymmetric linear sigma model.

---

*I thank Giorgio Parisi for this observation.

The formulation of a supersymmetric nonlinear sigma model with the symmetry-breaking pattern (3) appears at first sight problematical, for the following reason: Nonlinear sigma models describing the spontaneous breakdown of a symmetry group G to H normally have variables which live on the coset space G/H. Such a model can be made sypersymmetric only if this space is a Kähler manifold.[9] However, the space suggested by (3) is

$$\frac{G}{H} = \frac{U(n) \times U(n)}{U(n)} \approx U(n) \tag{7}$$

which is not even a complex manifold, and therefore is not Kähler. I choose to interpret this difficulty as a requirement from supersymmetry that there be additional light bosons in the theory beyond the required Goldstone bosons. The spectrum of these particles should be determined by embedding (7) in a larger space which is a Kähler manifold.[8] The smallest such homogeneous space with (7) as a subspace is

$$\frac{U(2n)}{U(n) \times U(n)} \, . \tag{8}$$

There are many embeddings of (7) into (8); for the purpose of this lecture, I will choose one and work out its implications. Let me, then, label the $U(n) \times U(n)$ subgroup of $U(2n)$ appearing in (8) as $[U(n) \times U(n)]_D$ and the isomorphic group appearing in (7) as $[U(n) \times U(n)]_N$. If I take $U(2n)$ in (8) to be generated by arbitrary Hermitian $2n \times 2n$ matrices, I can represent an embedding of (7) into (8) by identifying as the generators of the various subgroups of this $U(2n)$ matrices of the following forms:[10]

$$\begin{aligned} [U(x) \times U(n)]_D : &\quad \left(\begin{array}{c|c} t_1 & 0 \\ \hline 0 & t_2 \end{array}\right) \\ [U(n) \times U(n)]_N : &\quad V \left(\begin{array}{c|c} t_1 & 0 \\ \hline 0 & t_2 \end{array}\right) V^{-1} \\ U(n) : &\quad \left(\begin{array}{c|c} t_1 & 0 \\ \hline 0 & t_1 \end{array}\right) \end{aligned} \tag{9}$$

where t_1, t_2 are $n \times n$ Hermitian matrices and

$$V = \frac{1}{\sqrt{2}} \left(\begin{array}{c|c} 1 & 1 \\ \hline -1 & 1 \end{array}\right) \in SU(2) \tag{10}$$

$[U(n) \times U(n)]_D$ and $[U(n) \times U(n)]_N$ coincide precisely on the $U(n)$ subgroup generated by the last line of (9); this group will play the role of the conserved vector $U(n)$ of SSQCD.

The manifold (8) has $2n^2$ dimensions, so the number of light bosons in the model is doubled from the number of Goldstone bosons associated with the symmetry-breaking (3). The $2n^2$ coordinates form two adjoint representations of the vectorial $U(n)$. The particle spectrum of this model may be given a plausible physical interpretation as follows: Since the theory with fermionic matter fields alone must contain Goldstone bosons composed of two fermions and the theory with bosonic matter fields only should contain Goldstone bosons composed of two bosons, the full SSQCD should contain two light pseudoscalar mesons, both of which supersymmetry could well require to be massless. These mesons, with the quantum numbers of

$$(\psi_{ri} \cdot \psi_{\bar{r}j}) \quad \text{and} \quad (A_{ri} \cdot A_{\bar{r}j}) \,, \tag{11}$$

do form two adjoint representations of $U(n)$.

The most general $U(2n)$-invariant Lagrangian with coordinates in (8) has been constructed some time ago by Zumino[8] and Aoyama[11] ; it may be written:

$$\mathcal{L} = \int d^4\theta \, f_\pi^2 \, tr \, log \, (1 + A\bar{A}) \tag{12}$$

where A is an $n \times n$ complex matrix. Under an infinitesimal $U(2n)$ transformation

$$U = 1 + iT \,, \quad T = \left(\begin{array}{c|c} t_1 & t \\ \hline \bar{t} & t_2 \end{array} \right) \,. \tag{13}$$

A transforms according to

$$\delta A = i(At_2 - t_1 A) + t + A\bar{t}A \,. \tag{14}$$

In principle, one could add terms to (12) to break its symmetry explicitly to $[U(n) \times U(n)]_N$; however, I will study only the simplest kinetic energy term (12) here.

Equation (12) describes a theory with manifest supersymmetry and $2n^2$ massless boson-fermion pairs. However, this theory is not yet an acceptable one, because it does not satisfy the requirement (6) above. One might try to give masses to the particles of this theory by adding to (12) a symmetry-breaking F term of the form

$$\int d^2\theta \, tr(m\Delta(a)) \,. \tag{15}$$

However, this term produces $(mass)^2$ for the Goldstone bosons proportional to m^2, a signal that $< \psi_r \cdot \psi_{\bar{r}} >= 0$. This problem can only be remedied by adding to (12) an F term of zeroth order in m. There is no such term invariant under all of $U(2n)$, but one can find an F term invariant to $[U(n) \times U(n)]_N$. The generators of this subgroup can be rewritten from (9) in the form

$$T = \left(\begin{array}{c|c} t_a & t_b \\ \hline t_b & t_a \end{array} \right) , \tag{16}$$

where t_a and t_b are Hermitian. For these generators, (14) specializes to

$$\delta A = i[A, t_a] + t_b + A t_b A \, . \tag{17}$$

There is a unique F term constructed from A which is invariant to (17):

$$\int d^2\theta \; h f_\pi \cdot tr \; (tan^{-1} A) \tag{18}$$

where h is a constant. In addition, there is a unique structure which transforms linearly as an $(\bar{n}, n)$ under $[U(n) \times U(n)]_N$:

$$\int d^2\theta \; f_\pi \; tr \; m \left(\frac{i}{1 - iA} \right) . \tag{19}$$

The bosonic part of (18) plus (19) has the following form:

$$f_\pi F^y \; tr \; X_y \left(\frac{h}{1 + A^2} + \frac{1}{1 - iA} (-m) \frac{1}{1 - iA} \right) \tag{20}$$

where F is the auxiliary field associated with A and the X_y span a complete set of $n \times n$ matrices.

Let us first study the Lagrangian (12) plus (18), with $m = 0$. Eliminating F yields the potential energy:

$$V(A) = h^2 \; tr \left(\frac{1}{(1 + A^2)} \, (1 + A\bar{A}) \, \frac{1}{(1 + \bar{A}^2)} \, (1 + \bar{A} A) \right) . \tag{21}$$

If A is Hermitian, this $V(A) = h^2$; this choice gives the minimum of (21). That fact poses a severe problem for the theory: Supersymmetry is spontaneously broken. The minimum, though, does have some redeeming features. First, the space of minima of $V(A)$ is isomorphic to $U(n)$, so the vacuum degeneracy is that expected from (3). Secondly, if one attempts to give the Goldstone bosons mass by adding the mass term (19) and treating it only as a first-order perturbation of this theory, one finds a correction to the potential (for Hermitian A)

$$\Delta V = tr \; (m A^2) \tag{22}$$

which, properly, gives the Goldstone bosons $(mass)^2$ proportional to m.

The problem I have noted is, however, neatly resolved by a more careful examination of the full theory. If one takes $m \neq 0$, in (20), one can find a point at which the coefficient of F in this term vanishes by setting

$$\begin{aligned} 0 &= \frac{h}{1+A^2} - \frac{1}{1-iA}\, m\, \frac{1}{1-iA} \\ &= \frac{h}{1-iA}\left(\frac{1-iA}{1+iA} - \frac{m}{h}\right)\frac{1}{1-iA} \end{aligned} \quad . \tag{23}$$

Thus, by setting

$$A = -i\left(\frac{1-m/h}{1+m/h}\right) \tag{24}$$

one finds a supersymmetric vacuum state. Note that if m is a matrix with one large eigenvalue, the corresponding eigenvalue of A is drawn to the value (24); this verifies the decoupling requirement (8) above.

This supersymmetric vacuum states exists for any $m \neq 0$, but as $m \to 0$ it is separated from the set of states with A Hermitian by a potential energy barrier whose height grows as m^{-1}. If we speak in terms of the index of Witten[3], the model I have constructed has index equal to 1 for any nonzero value of m but a zero index at $m = 0$. This discontinuous change in the index at $m = 0$, or, equivalently, the inaccessibility of the supersymmetric state as $m \to 0$, violates an explicit assumption made by Witten in extending his proof the absence of spontaneous sypersymmetry breaking in SSQCD to the massless case.[3] This method of evading Witten's conclusion was suggested earlier by Srednicki;[12] I thought at the time that it could never be realized in an explicit model of SSQCD.

I have, then, presented an effective Lagrangian which describes the low-energy of dynamics of supersymmetric Yang-Mills theory with matter fields in complex-conjugate-pair representations, assuming that the pattern of chiral symmetry breaking is that observed in the familiar strong interactions. This Lagrangian has a supersymmetric vacuum state for any nonzero value of the matter field mass, but it has spontaneously broken supersymmetry for matter fields of precisely zero mass.

One can straightforwardly extend the analysis of this paper to more general forms of the non-linear sigma model action and to the case of matter fields in real representations. In all cases, the physics of the generalized models is qualitatively the same as that described here.[13]

I am grateful to Chong-Leong Ong, Giorgio Parisi, Gabriele Veneziano, Edward Witten, and Shimon Yankielowicz, for discussions of the properties of SSQCD, to I. M. Singer for valuable advice on geometry, and, especially, to Joe Polchinski, for asking all the right questions. I thank Glennys Farrar and Frank Henyey for organizing this stimulating conference.

REFERENCES

1. M. Dine, W. Fischler, and M. Srednicki, Nucl. Phys. B189, 575 (1981).
2. S. Dimopoulos and S. Raby, Nucl. Phys. B192, 353 (1981).
3. E. Witten, Nucl. Phys. B202, 513 (1981).
4. T. R. Taylor, G. Veneziano, and S. Yankielowicz, CERN preprint TH 3460 (1980).
5. S. Weinberg, Phys. Rev. Lett. 29, 1698 (1967), Phys. Rev. 166, 1568 (1968).
6. T. Banks and A. Casher, Nucl. Phys. B169, 102 (1980).
7. B. A. Ovrut and J. Wess, Phys. Rev. D25, 409 (1982).
8. C.-L. Ong, SLAC preprint SLAC-PUB-2956 (1982).
9. B. Zumino, Phys. Lett. 87B, 203 (1979).
10. I. M. Singer, personal communication.
11. S. Aoyama, Nuov. Cim. 57A, 176 (1980).
12. M. Srednicki, personal communication.
13. M. E. Peskin, in preparation.

Light-Cone Superfields for Extended Supergravity in Ten Dimensions and Type II Superstrings*

JOHN H. SCHWARZ

California Institute of Technology, Pasadena CA 91125

ABSTRACT

A new extended supergravity theory in ten dimensions is formulated in terms of an unconstrained scalar light-cone superfield. Since this theory describes the massless sector of type II superstring theory, the formalism is then extended to fields that are functionals of string superspace coordinates. In this way a field theory of superstrings is obtained. The theory is expected to be nonpolynomial, but so far it has only been constructed to order κ.

Introduction

Supersymmetrical string theories (or "superstring" theories) have been proposed for use as a fundamental theory of elementary particles unifying the description of gravitation with the other interactions [1]. This program has advanced rapidly in the last few years with the development by Michael Green and myself of a formalism in which the spacetime supersymmetry of the theories is revealed [2]. The most recent developments, described here in abridged form, include the construction of a new extended supergravity theory in ten dimensions that corresponds to the massless sector of type II superstring theory [3] and a field-theoretic description of the superstrings themselves based on a superfield that is a functional of light-cone superspace coordinates [4,5].

Conventional quantum field theories based on a finite number of point-particle fields (including supergravity) appear incapable of providing a perturbatively finite or renormalizable theory of gravitation.The outlook for superstring theories is more sanguine, even though no proofs have been given yet beyond one loop. Superstring theories contain two fundamental parameters, a coupling constant κ and a Regge-slope parameter α' that is inversely proportional to the string tension. (The combination $\kappa(\alpha')^{-2}$ is dimensionless.) In addition, there is a length scale R characterizing the size of six compact dimensions, required for realism since the theories are altogether ten-dimensional. It is hoped that R will be related to α' and κ when one succeeds in finding classical solutions that describe the compactification. This step should also be responsible for

* Work supported in part by the US Department of Energy under contract No. DE-AC03-81-ER40050.

supersymmetry breaking and the introduction of Yang-Mills gauge symmetries. It will become feasible to explore nontrivial compactifications when the full nonlinear field theory is explicitly constructed, but so far only the order κ interactions have been described in the appropriate field-theory language.

Both type I and type II superstring theories are free from ghosts and tachyons and require ten-dimensional spacetime for their consistency as quantum theories. Type I theories have one ten-dimensional supersymmetry and describe unoriented open and closed strings. They allow orthogonal or symplectic Yang-Mills gauge groups to be introduced and reduce to N=4 Yang-Mills field theory in a suitable α', $R \to 0$ limit. It has been shown at one loop that the only divergences occurring in the perturbation expansion of type I theories can be removed by renormalization of α'. Theory II has two ten-dimensional supersymmetries and describes oriented closed strings only. It is a unique theory, with no freedom to choose a gauge group, and reduces to N=8 supergravity in a suitable α', $R \to 0$ limit. The evidence from one-loop and topological considerations suggests that theory II is finite to all orders in perturbation theory, even though the limiting N=8 supergravity theory probably has ultraviolet divergences starting at three loops.

The purpose of this talk is to sketch the field-theoretic description of type II superstring theory. In older string theories one started from a covariant gauge-invariant formulation and used the gauge invariance to choose a light-cone gauge [6]. In the case of type II superstrings, as well as the field theory that describes the massless sector, there seems to be no manifestly covariant action principle, so it is necessary to use a physical-gauge formulation from the outset [7]. I will first describe the field theory (to order κ) and then indicate how the extension to the string theory works. Both cases use a "light-cone superspace" description involving eight anticommuting Grassmann coordinates ϑ^a that transform as a spinor under the SO(8) that rotates the eight transverse directions. In generalizing to the string, ϑ^a and the eight transverse coordinates x^i become functions of σ, a parameter that labels the points along the length of the string. Also, some of the super-Poincaré symmetries are realized locally in σ, while others are not. The locally conserved quantities correspond physically to continuity of $x^i(\sigma)$ and $\vartheta^a(\sigma)$ when strings interact, as well as local conservation of the conjugate momentum densities $p^i(\sigma)$ and $\lambda^a(\sigma)$.

Extended Supergravity in Ten Dimensions

There are two inequivalent extended supergravity theories in ten dimensions, both of which correspond to N = 8 supergravity on truncation to four dimensions. The physical states of the two theories belong to different multiplets of SO(8), the group whose representations label massless states in ten dimensions, although they coincide on truncation to nine or fewer dimensions. One theory, containing two inequivalent gravitino representations, can be obtained by truncation of d = 11 supergravity. The other, to be considered here, contains two equivalent gravitinos and cannot be obtained from a higher dimension. Besides the graviton and gravitinos it contains a pair of spinors, scalars, and second-rank antisymmetric tensors, as well as a single self-dual fourth-rank antisymmetric tensor. This is precisely the particle content described by a scalar superfield $\Phi(x,\vartheta)$, where ϑ^a is an 8-component spinor Grassmann coordinate.

The spacetime manifold is described by light-cone coordinates $x^{\pm} = (x^0 \pm x^9)/\sqrt{2}$ and transverse coordinates x^i, $i = 1,2,\ldots,8$. The theory is conveniently described in a Hamiltonian formalism with H regarded as the generator of translations in x^+, the light-cone "time."The other coordinates are Fourier transformed so that

$$\alpha \equiv 2p^+ = 2i \frac{\partial}{\partial x^-} \tag{1}$$

$$p^i = -i \frac{\partial}{\partial x^i} \tag{2}$$

$$\lambda^a = \frac{\partial}{\partial \vartheta^a} \ , \tag{3}$$

and we use a momentum-space field $\Phi(x^+, \alpha, p, \lambda)$. The expansion of this field in powers of λ contains 128 Bose modes and 128 Fermi modes and has its coefficients arranged so that its hermitian conjugate equals its Grassmann Fourier transform

$$\Phi^{\dagger}(\alpha,p,\lambda) = \frac{\alpha^4}{16} \int \Phi\,(\alpha,p,\lambda')\, e^{2\lambda\lambda'/\alpha} d^8\lambda'. \tag{4}$$

This property is crucial for demonstrating hermiticity of the interactions [3].

The algebra of SO(8) involves eight matrices $\gamma^i_{a\dot{a}}$, where dotted and undotted spinor indices refer to inequivalent 8's of SO(8). These matrices satisfy the usual Dirac algebra. (An explicit representation is given in Appendix A of ref. [4].) A supersymmetry charge in ten dimensions has 16 real components that decompose under transverse SO(8) into two eight-component spinors q^{+a} and $q^{-\dot{a}}$. For the extended supergravity theory there are two of each, represented in terms of superspace coordinates and momenta as follows:

$$q_1^{+a} = \frac{\alpha}{\sqrt{2}} \vartheta^a \qquad\qquad q_1^{-\dot{a}} = (\gamma\cdot p\,\vartheta)^{\dot{a}} \tag{5a}$$

$$q_2^{+a} = \sqrt{2}\,\lambda^a \qquad\qquad q_2^{-\dot{a}} = \frac{2}{\alpha}(\gamma\cdot p\,\lambda)^{\dot{a}}. \tag{5b}$$

These satisfy, in particular,

$$\{ q_1^{-\dot{a}}, q_2^{-\dot{b}} \} = 2h\,\delta^{\dot{a}\dot{b}}, \tag{6}$$

where

$$h = p^2/\alpha \tag{7}$$

plays the role of the Hamiltonian, since the mass-shell condition is $h = p^-$. The Lorentz generators are represented in similar fashion:

$$j^{i+} = \frac{1}{2} x^i \alpha - x^+ p^i \tag{8a}$$

$$j^{ij} = x^i p^j - x^j p^i - \frac{i}{2} \vartheta \gamma^{ij} \lambda \tag{8b}$$

$$j^{+-} = x^+ h + i \alpha \frac{\partial}{\partial \alpha} - \frac{i}{2} \vartheta \lambda + \frac{5i}{2} \tag{8c}$$

$$j^{i-} = x^i h + 2i p^i \frac{\partial}{\partial \alpha} - \frac{i}{\alpha} \vartheta \gamma^i \gamma \cdot p \lambda + \frac{3ip^i}{\alpha}. \tag{8d}$$

When any generator is represented in terms of fields, it is denoted by the corresponding capital letter. In general for the free theory

$$G = \frac{1}{2} \int \alpha \Phi(-\alpha, -p, -\lambda)\, g\, \Phi(\alpha, p, \lambda)\, d\alpha\, d^8 p\, d^8 \lambda, \tag{9}$$

as is easily verified using the equal x^+ commutation relation

$$[\, \Phi(1), \Phi(2)\,] = \frac{1}{\alpha_1} \delta(\alpha_1 + \alpha_2)\, \delta^8(p_1 + p_2)\, \delta^8(\lambda_1 + \lambda_2). \tag{10}$$

In the interacting field theory this form of G is exact for P^i, J^{i+}, J^{ij}, Q_1^{+a}, and Q_2^{+a}, but $J^{+-}, J^{i-}, H, Q_1^{-\dot{a}}$, and $Q_2^{-\dot{a}}$ include additional interaction terms involving cubic and higher powers of Φ. The Hamiltonian, in particular, has an expansion

$$H = H_2 + \kappa H_3 + \kappa^2 H_4 + \cdots, \tag{11}$$

where H_2 has the form given in eq. (9).

The cubic contribution to H and the other generators, and undoubtedly the higher-order terms as well, are determined by the super-Poincaré algebra. The result is unique if one demands a correspondence with the massless sector of the string theory described in the next section. We find that

$$H_3 = \int d\mu_3^0\, \mathbb{P}^i \mathbb{P}^j v^{ij}(\Lambda) \Phi(1)\Phi(2)\Phi(3) \tag{12}$$

$$Q^{-\dot{a}}_{\text{cubic}} = \frac{1}{\sqrt{2}} \left[\eta\, Q_1^{-\dot{a}} + \eta^* Q_2^{-\dot{a}}\right]_{\text{cubic}}$$

$$= \int d\mu_3^0\, \mathbb{P}^i s^{i\dot{a}}(\Lambda)\, \Phi(1)\Phi(2)\Phi(3) \tag{13a}$$

$$\widetilde{Q}^{-\dot{a}}_{\text{cubic}} = \frac{1}{\sqrt{2}} \left[\eta^* Q_1^{-\dot{a}} + \eta\, Q_2^{-\dot{a}}\right]_{\text{cubic}}$$

$$= \int d\mu_3^0\, \mathbb{P}^i \widetilde{s}^{i\dot{a}}(\Lambda) \Phi(1)\Phi(2)\Phi(3)\ , \tag{13b}$$

where

$$\eta = e^{i\pi/4} \tag{14}$$

$$\mathbb{P}^i = \alpha_1 p_2^i - \alpha_2 p_1^i \tag{15}$$

$$\Lambda^{a} = \alpha_1 \lambda_2^{a} - \alpha_2 \lambda_1^{a} \tag{16}$$

$$d\mu_3^0 = \Big(\prod_{r=1}^{3} d\alpha_r\, d^8 p_r d^8 \lambda_r \Big) \delta(\Sigma \alpha_r) \delta^8(\Sigma p_r) \delta^8(\Sigma \lambda_r). \tag{17}$$

The quantities P^i and Λ^a effectively have total antisymmetry in the labels 1, 2, 3 when the δ functions in the measure are taken into account.Also, the functions v^{ij}, $s^{i\dot{a}}$, and $\tilde{s}^{i\dot{a}}$ are polynomials in Λ uniquely determined by the normalization condition $v^{ij}(0) = \delta^{ij}$ and the equations

$$\Big(\eta \Lambda^a + \frac{\alpha}{2} \eta^* \frac{\partial}{\partial \Lambda^a} \Big) v^{ij}(\Lambda) = \frac{i}{\sqrt{2}} \gamma^j_{a\dot{a}} s^{i\dot{a}}(\Lambda) \tag{18a}$$

$$\Big(\eta^* \Lambda^a + \frac{\alpha}{2} \eta \frac{\partial}{\partial \Lambda^a} \Big) v^{ij}(\Lambda) = -\frac{i}{\sqrt{2}} \gamma^i_{a\dot{a}} \tilde{s}^{j\dot{a}}(\Lambda). \tag{18b}$$

These equations are derived by requiring that the anticommutation rules

$$\{ Q^{-\dot{a}}, Q^{-\dot{b}} \} = \{ \tilde{Q}^{-\dot{a}}, \tilde{Q}^{-\dot{b}} \} = 2H\, \delta^{\dot{a}\dot{b}} \tag{19}$$

$$\{ Q^{-\dot{a}}, \tilde{Q}^{-\dot{b}} \} = 0 \tag{20}$$

are satisfied to order κ*. We expect the supersymmetry algebra to uniquely determine the higher-order terms as well, but the mathematics is somewhat complicated and has not yet been completed. The N = 8 theory in four dimensions is obtained by a trivial truncation, i.e., by dropping the dependence of the field on six of the eight transverse coordinates. The resulting emergence of an SU(8) symmetry is described in ref. [3].

Type II Superstrings

Type II superstrings are described by light-cone superspace coordinates $x^i(\sigma)$ and $\vartheta^a(\sigma)$ in position space or $p^i(\sigma)$ and $\lambda^a(\sigma)$ in momentum space, where σ is a parameter labeling points on the string in such a way that the density of p^+ is constant. Accordingly, the range of σ must be proportional to the total p^+ of the string, and is taken to be $-\pi|\alpha| \leq \sigma \leq \pi|\alpha|$. Since type II strings are closed it is convenient to extend the range of σ, imposing the periodicity requirements

$$x^i(\sigma) = x^i(\sigma + 2\pi\alpha) \tag{21a}$$

$$\vartheta^a(\sigma) = \vartheta^a(\sigma + 2\pi\alpha), \tag{21b}$$

with similar relations for $p^i(\sigma)$ and $\lambda^a(\sigma)$. The field that creates or destroys a string is the scalar functional $\Phi[x^+, \alpha, p(\sigma), \lambda(\sigma)]$, subject to the constraint

* Actually, this only gives the ij symmetrical parts of eqs. (18), which by themselves do not have a unique solution. The ambiguity has been resolved by requiring a string generalization. A different choice was made in ref. [3].

$$\Phi[p(\sigma), \lambda(\sigma)] = \Phi[p(\sigma+\sigma_0), \lambda(\sigma+\sigma_0)]. \tag{22}$$

This corresponds to a rigid displacement of the parametrization along the string, which clearly should have no physical significance. It is the only reparametrization symmetry present in the light-cone formalism.

The superstring field theory can be written succinctly in functional language. However, this conceals a number of subtle issues and may not be the most useful form for explicit calculations. In any case we have found it essential to insert explicit mode expansions and to study the interaction of specific string modes.In this way one always deals with unambiguous and well-defined expressions, which in the end can be reassembled into functional expressions. For this purpose the coordinates are expanded in Fourier series as follows (in units with $\alpha' = 1$):

$$x^i(\sigma) = x^i + \sum_{n\neq 0} \frac{1}{n}\left(\alpha_n^i e^{in\sigma/|\alpha|} + \tilde{\alpha}_n^i e^{-in\sigma/|\alpha|}\right) \tag{23}$$

$$p^i(\sigma) = \frac{1}{2\pi|\alpha|}[\, p^i + \frac{1}{2}\sum_{n\neq 0} (\alpha_n^i e^{in\sigma/|\alpha|} + \tilde{\alpha}_n^i e^{-in\sigma/|\alpha|})\,] \tag{24}$$

$$q^{+a}(\sigma) = \eta^* \lambda^a(\sigma) + \frac{e(\alpha)}{4\pi}\eta\vartheta^a(\sigma) = \frac{1}{2\pi|\alpha|}\sum_{-\infty}^{\infty} Q_n^a e^{in\sigma/|\alpha|} \tag{25a}$$

$$\tilde{q}^{+a}(\sigma) = \eta\lambda^a(\sigma) + \frac{e(\alpha)}{4\pi}\eta^*\vartheta^a(\sigma) = \frac{1}{2\pi|\alpha|}\sum_{-\infty}^{\infty} \tilde{Q}_n^a e^{in\sigma/|\alpha|}\ , \tag{25b}$$

where

$$e(\alpha) \equiv \frac{\alpha}{|\alpha|}\ , \tag{26}$$

and the oscillators have the commutation relations

$$[\,\alpha_m^i, \alpha_n^j\,] = [\,\tilde{\alpha}_m^i, \tilde{\alpha}_n^j\,] = m\,\delta^{ij}\,\delta_{m+n,0} \tag{27}$$

$$\{\,Q_m^a, Q_n^b\,\} = \{\,\tilde{Q}_m^a, \tilde{Q}_n^b\,\} = \alpha\,\delta^{ab}\delta_{m+n,0} \tag{28}$$

$$[\,\alpha_m^i, \tilde{\alpha}_n^j\,] = \{\,Q_m^a, \tilde{Q}_n^b\,\} = 0, \tag{29}$$

so that

$$[\,x^i(\sigma), p^j(\sigma')\,] = i\,\delta^{ij}\delta(\sigma-\sigma') \tag{30}$$

$$\{\,\vartheta^a(\sigma), \lambda^b(\sigma')\,\} = \delta^{ab}\delta(\sigma-\sigma') \tag{31}$$

$$[\,x^i(\sigma), x^j(\sigma')\,] = [\,p^i(\sigma), p^j(\sigma')\,] = 0 \tag{32}$$

$$\{\,\vartheta^a(\sigma), \vartheta^b(\sigma')\,\} = \{\,\lambda^a(\sigma), \lambda^b(\sigma')\,\} = 0. \tag{33}$$

Note that positive subscripts represent lowering operators and negative subscripts give the corresponding raising operators.

The Hamiltonian generalizing eq. (7) is given by

$$h = \int_{-\pi|\alpha|}^{\pi|\alpha|} \left[2\pi e(\alpha) p^2 + \frac{e(\alpha)}{8\pi}[x'(\sigma)]^2 + 2\pi \lambda'(\sigma)\lambda(\sigma) - \frac{1}{8\pi}\vartheta'(\sigma)\vartheta(\sigma) \right] d\sigma$$

$$= \frac{1}{\alpha}(p^2 + N + \tilde{N}), \tag{34}$$

where

$$N = \sum_{n=1}^{\infty} \left(\alpha_{-n}^i \alpha_n^i + \frac{n}{\alpha} Q_{-n}^a Q_n^a \right) \tag{35a}$$

$$\tilde{N} = \sum_{n=1}^{\infty} \left(\tilde{\alpha}_{-n}^i \tilde{\alpha}_n^i + \frac{n}{\alpha} \tilde{Q}_{-n}^a \tilde{Q}_n^a \right). \tag{35b}$$

In terms of these operators the constraint in eq. (22) takes the form $(N - \tilde{N})\Phi = 0$. Normal ordering the oscillators in the integrand of eq. (34) gives (infinite) constants that cancel between the bosonic and fermionic terms, so the expression is correct as written. This feature is one of many advantages of superstrings over older nonsupersymmetrical string models.

The supersymmetries in eq. (25) are locally conserved (as functions of σ) in string interactions. This means that $e(\alpha)\vartheta^a(\sigma)$ and the Grassmann momentum $\lambda^a(\sigma)$ are conserved. The former can be interpreted as expressing continuity of the coordinate between initial and final string configurations. ($e(\alpha)$ is positive for incoming strings and negative for outgoing ones.) In similar fashion $e(\alpha)x^i(\sigma)$ and the ordinary momentum density $p^i(\sigma)$ are also conserved locally. These may also be interpreted as consequences of local conservation of the j^{i+} density

$$j^{i+}(\sigma) = \frac{1}{\pi} e(\alpha) x^i(\sigma) - x^+ p^i(\sigma) \ . \tag{36}$$

The q^- generators, on the other hand, describe global symmetries only. The formulas are

$$q_1^- = \int_{-\pi|\alpha|}^{\pi|\alpha|} [\, \gamma\cdot p(\sigma)\vartheta(\sigma) + ie(\alpha)\gamma\cdot x'(\sigma)\lambda(\sigma)]\, d\sigma \tag{37a}$$

$$q_2^- = \int_{-\pi|\alpha|}^{\pi|\alpha|} [\, 4\pi e(\alpha)\gamma\cdot p(\sigma)\lambda(\sigma) - \frac{i}{4\pi}\gamma\cdot x'(\sigma)\vartheta(\sigma)]\, d\sigma \tag{37b}$$

or

$$q^- = \frac{1}{\sqrt{2}}(\eta q_1^- + \eta^* q_2^-) = \frac{\sqrt{2}}{\alpha}\sum_{-\infty}^{\infty} \gamma\cdot\alpha_{-n} Q_n \tag{38a}$$

$$\tilde{q}^- = \frac{1}{\sqrt{2}}(\eta^* q_1^- + \eta q_2^-) = \frac{\sqrt{2}}{\alpha}\sum_{-\infty}^{\infty} \gamma\cdot\tilde{\alpha}_{-n} \tilde{Q}_n . \tag{38b}$$

The passage to representations of the algebra in terms of fields works essentially the same as in the field theory with

$$G = \frac{1}{2} \int \alpha \, \Phi[-\alpha, -p(\sigma), -\lambda(\sigma)] \; g \; \Phi[\alpha, p(\sigma), \lambda(\sigma)] d\alpha \, D^8 p(\sigma) D^8 \lambda(\sigma) \quad (39)$$

replacing eq. (9) and

$$[\Phi(1), \Phi(2)] = \frac{1}{\alpha_1} \delta(\alpha_1 + \alpha_2) \, \Delta^8[p_1(\sigma) + p_2(\sigma)] \, \Delta^8[\lambda_1(\sigma) + \lambda_2(\sigma)] \quad (40)$$

replacing eq. (10). The functional Δ functions and functional integrations are given a precise meaning as infinite products over the sine and cosine modes of the Fourier expansions.

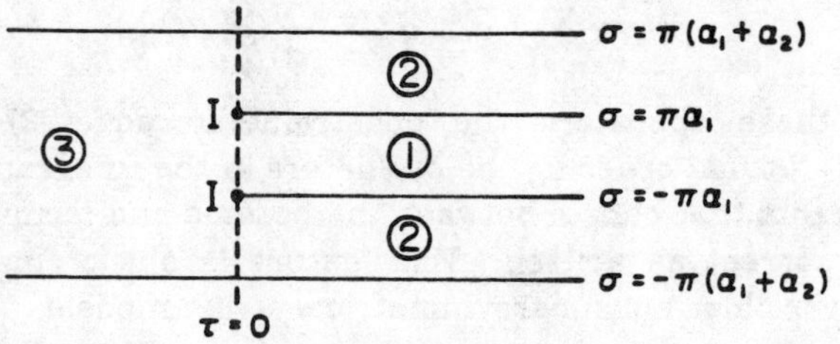

Strings #1 and #2 join at I to form string #3. The boundaries of regions 1, 2, and 3 are identified so as to describe closed strings.

Consider now the three-string interaction shown in fig. 1, in which incoming strings #1 and #2 join to form outgoing string #3. A parameter σ_r is associated with string r according to the rule

$$\sigma_1 = \sigma \qquad -\pi\sigma_1 \leq \sigma \leq \pi\alpha_1 \quad (41a)$$

$$\sigma_2 = \begin{cases} \sigma - \pi\alpha_1 & \pi\alpha_1 \leq \sigma \leq \pi(\alpha_1 + \alpha_2) \\ \sigma + \pi\alpha_1 & -\pi(\alpha_1 + \alpha_2) \leq \sigma \leq \pi(\alpha_1 + \alpha_2) \end{cases} \quad (41b)$$

$$\sigma_3 = -\sigma \qquad -\pi(\alpha_1 + \alpha_2) \leq \sigma \leq \pi(\alpha_1 + \alpha_2). \quad (41c)$$

Then the momentum densities are given by

$$p_r^i(\sigma) = \Theta_r \, p_r^i(\sigma_r) \quad (42)$$

$$\lambda_r^a(\sigma) = \Theta_r \, \lambda_r^a(\sigma_r), \quad (43)$$

where the Θ_r's are the step functions

$$\Theta_1 = \vartheta(\pi\alpha_1 - |\sigma|) \quad (44a)$$

$$\Theta_2 = \vartheta(|\sigma| - \pi\alpha_1) \tag{44b}$$

$$\Theta_3 = \Theta_1 + \Theta_2 = 1. \tag{44c}$$

With these definitions and analogous ones for $x_r(\sigma)$ and $\vartheta_r(\sigma)$ the local conservation laws are

$$\sum_{r=1}^{3} p_r^i(\sigma) = \sum_{r=1}^{3} \lambda_r^a(\sigma) = 0 \tag{45}$$

$$\sum_{r=1}^{3} e\,(\alpha_r) x_r^i(\sigma) = \sum_{r=1}^{3} e\,(\alpha_r)\vartheta_r^a(\sigma) = 0. \tag{46}$$

The cubic interaction Hamiltonian is written in the form

$$H_3 = \int \mathrm{d}\mu_3\, G\, \Phi(1)\Phi(2)\Phi(3), \tag{47}$$

where

$$\mathrm{d}\mu_3 = \left[\prod_{r=1}^{3} \mathrm{d}\alpha_r D^8\lambda_r(\sigma) D^8 p_r(\sigma)\right] \delta(\Sigma\alpha_r)\, \Delta^8[\Sigma p_r(\sigma)]\, \Delta^8[\Sigma\lambda_r(\sigma)], \tag{48}$$

and G is an operator to be determined. The local conservation laws in eqs. (45) and (46) are satisfied as a consequence of the Δ functionals in eq. (48) provided that G is of suitable form. The string fields Φ may be regarded as infinite component fields containing an ordinary point-particle field for every set $\{n\}$ of excitations of the various Bose and Fermi oscillators (subject to the constraint of eq. (22)). Therefore the information contained in eq. (47) is equivalent to an infinite set of couplings $C(\{n^{(1)}, n^{(2)}, n^{(3)}\})$ describing the interaction of three arbitrary string states. This information is conveniently collected in a giant ket vector

$$|V\rangle = \sum_{\{n^{(1)}, n^{(2)}, n^{(3)}\}} C(\{n^{(1)}, n^{(2)}, n^{(3)}\})\, |\{n^{(1)}, n^{(2)}, n^{(3)}\}\rangle, \tag{49}$$

where we have associated a Fock-space state with each set of mode numbers. This form of the vertex can be calculated by inserting explicit mode expansions for the fields in eq. (47) and doing the integrals, all of which are Gaussians. The result has the form

$$|V\rangle = G_{op}\, E_\alpha E_Q\, |0\rangle \delta(\Sigma\alpha_r)\delta^8(\Sigma p_r)\delta^8(\Sigma\lambda_r), \tag{50}$$

where G_{op} corresponds to G in eq. (47), and

$$E_\alpha = \exp\{\, \frac{1}{2}\Sigma(\, \alpha_{-m}^{(r)} \bar{N}_{mn}^{rs} \alpha_{-n}^{(s)} + \tilde{\alpha}_{-m}^{r}\, \bar{N}_{mn}^{rs} \tilde{\alpha}_{-n}^{s}\,)$$

$$+ \mathbb{P} \sum \bar{N}_m^r (\alpha_{-m}^{(r)} + \tilde{\alpha}_{-m}^{(r)}) - \frac{\tau_0}{\alpha} \mathbb{P}^2 \} \tag{51}$$

$$E_Q = \exp\{\ \frac{1}{2}\sum \frac{m}{\alpha_r}\,(\,Q^{(r)}_{-m}\bar{N}^{rs}_{mn}\,Q^{(s)}_{-n} + \tilde{Q}^{(r)}_{-m}\bar{N}^{rs}_{mn}\,\tilde{Q}^{(s)}_{-n})$$

$$+\ \frac{i}{2}\alpha(\sum Q^{(r)}_{-m}\frac{m}{\alpha_r}\bar{N}^{r}_{m})(\sum \tilde{Q}^{(s)}_{-n}\frac{n}{\alpha_s}\bar{N}^{s}_{m}) - \Lambda\sum \frac{m}{\alpha_r}\bar{N}^{r}_{m}(\eta^*Q^{(r)}_{-m} + \eta\tilde{Q}^{(r)}_{-m})\}. \quad (52)$$

In these expressions

$$\alpha\ = \alpha_1\alpha_2\alpha_3 \quad (53)$$

$$\tau_0\ = \sum_{r=1}^{3} \alpha_r \ln|\alpha_r| \quad (54)$$

$$\bar{N}^{rs}_{mn}\ = -\ \frac{mn\,\alpha}{n\alpha_r + m\alpha_s}\bar{N}^{r}_{m}\bar{N}^{s}_{n} \quad (55)$$

$$\bar{N}^{r}_{m}\ = \frac{1}{\alpha_r} f_m(-\frac{\alpha_{r+1}}{\alpha_r})e^{m\tau_0/\alpha_r} \quad (56)$$

$$f_m(\gamma) = \frac{1}{m!}\,\frac{\Gamma(m\gamma)}{\Gamma(m\gamma+1-m)}\ , \quad (57)$$

and $\mathbb{P}$ and Γ are as given in eqs. (15) and (16). Using various identities among these quantities one can show that the sums in eqs. (45) and (46) annihilate $|V>$ provided that they commute with G_{op}.

The next step is to determine G_{op} in eq. (50). The first requirement is that it should commute with the quantities in eqs. (45) and (46). This is satisfied by the three linear expressions

$$X = \mathbb{P} - \sum_{r=1}^{3}\sum_{n=1}^{\infty}\frac{n\alpha}{\alpha_r}\bar{N}^{r}_{n}\,\alpha^{(r)}_{-n} \quad (58a)$$

$$\tilde{X} = \mathbb{P} - \sum_{r=1}^{3}\sum_{n=1}^{\infty}\frac{n\alpha}{\alpha_r}\bar{N}^{r}_{n}\,\tilde{\alpha}^{(r)}_{-n} \quad (58b)$$

$$Y = \Lambda - \frac{1}{2}\sum_{r=1}^{3}\sum_{n=1}^{\infty}\frac{n\alpha}{\alpha_r}\bar{N}^{r}_{n}\left[Q^{(r)}_{-n} + \tilde{Q}^{(r)}_{-n}\right]. \quad (59)$$

There is one slightly subtle case, namely

$$[\,X - \tilde{X},\ \sum p_r(\sigma)\,] \propto \delta(\sigma - \pi\alpha_1) - \delta(\sigma + \pi\alpha_1)\ , \quad (60)$$

which appears to violate one of the conditions. However, $\sigma = \pi\alpha_1$ and $\sigma = -\pi\alpha_1$ correspond to the same spacetime point, i.e., the place where the interaction occurs, and therefore the total momentum density at the interaction point is conserved.We now rewrite eq. (50) and the cubic supersymmetry operators in the form

$$|V> = \tilde{X}^i X^j v^{ij}(Y)E_\alpha E_Q|0>\,\delta^{17} \quad (61)$$

$$|Q^{-\dot{a}}> = \widetilde{X}^i s^{i\dot{a}}(Y) E_\alpha E_Q |0> \delta^{17} \tag{62a}$$

$$|\widetilde{Q}^{-\dot{a}}> = X^i \widetilde{s}^{i\dot{a}}(Y) E_\alpha E_Q |0> \delta^{17}, \tag{62b}$$

which automatically builds in the local symmetry requirements and reduces the problem to the determination of functions $v^{ij}, \widetilde{s}^{i\dot{a}}$, and $s^{i\dot{a}}$. Implementing the supersymmetry algebra of eq. (19) to order κ leads to eqs. (18a and b) with Λ replaced by Y. Equation (20) in order κ then serves as a nontrivial consistency check on the solution. Explicit expansions of $v, \widetilde{s}$, and s are given in Appendix D of ref. [5].

The functional equivalents of these oscillator results can now be described. The basic idea is that there are derivative couplings at the interaction point $\sigma = \pm \pi\alpha_1$. However, these are singular and need to be described as limits. Thus, we find by explicit comparison with the oscillator expressions that

$$H_3 = \lim_{\sigma \to \pi\alpha_1} \int d\mu_3 \widetilde{X}^i(\sigma) X^j(\sigma) v^{ij}(Y(\sigma)) \Phi(1)\Phi(2)\Phi(3) \tag{63}$$

$$Q^{-\dot{a}}_{\text{cubic}} = \lim_{\sigma \to \pi\alpha_1} \int d\mu_3 \widetilde{X}^i(\sigma) s^{i\dot{a}}(Y(\sigma)) \Phi(1)\Phi(2)\Phi(3) \tag{64a}$$

$$\widetilde{Q}^{-\dot{a}}_{\text{cubic}} = \lim_{\sigma \to \pi\alpha_1} \int d\mu_3 X^i(\sigma) \widetilde{s}^{i\dot{a}}(Y(\sigma)) \Phi(1)\Phi(2)\Phi(3), \tag{64b}$$

where

$$X(\sigma) = -2\pi\sqrt{-\alpha}\,(\pi\alpha_1 - \sigma)^{1/2}[p^{(1)}(\sigma) - \frac{1}{4\pi} x^{(1)\prime}(\sigma)$$
$$+ p^{(1)}(-\sigma) - \frac{1}{4\pi} x^{(1)\prime}(-\sigma)] \tag{65a}$$

$$\widetilde{X}(\sigma) = -2\pi\sqrt{-\alpha}(\pi\alpha_1 - \sigma)^{1/2}[p^{(1)}(\sigma) + \frac{1}{4\pi} x^{(1)\prime}(\sigma)$$
$$+ p^{(1)}(-\sigma) + \frac{1}{4\pi} x^{(1)\prime}(-\sigma)] \tag{65b}$$

$$Y(\sigma) = -2\pi\sqrt{-\alpha}\,(\pi\alpha_1 - \sigma)^{1/2}[\lambda^{(1)}(\sigma) + \lambda^{(1)}(-\sigma)]. \tag{66}$$

The vanishing factors in these expressions combine with the singular behavior of the operations to give the desired result in the limit. Equations (65) and (66) are expressed in terms of the variables of string # 1, but in the limit the choice of any of the strings is equivalent, and the vertex in fact has total symmetry in the three strings.

It is important to complete the construction of the field theory of sect. 2 and the string theory of this section. Both are expected to be nonpolynomial. Once this is achieved, we should be in a position to look for classical solutions that describe physically interesting compactifications and to study the quantum behavior of the theories more completely than has been possible with previous formalisms. The prospects are quite exciting.

References

[1] J. Scherk and J. H. Schwarz, Nucl. Phys. B81 (1974) 118; Phys. Lett. 57B (1975) 463; J. H. Schwarz in "New Frontiers in High-Energy Physics," Proc. Orbis Scientiae 1978, ed. A. Perlmutter and L. F. Scott (Plenum, New York, 1978) p. 431.

[2] J. H. Schwarz "Superstring Theory," Phys. Reports Vol. 89, No. 3, 1982 and references therein.

[3] M. B. Green and J. H. Schwarz, Caltech preprint CALT-68-957, to be published in Phys. Lett. B.

[4] M. B. Green and J. H. Schwarz, Caltech preprint CALT-68-956, to be published in Nucl. Phys. B.

[5] M. B. Green, J. H. Schwarz, and L. Brink, Caltech preprint CALT-68-972.

[6] P. Goddard, J. Goldstone, C. Rebbi, and C. B. Thorn, Nucl. Phys. B56 (1973) 109.

[7] N. Marcus and J. H. Schwarz, Phys. Lett. 115B (1982) 111.

PHASE TRANSITIONS AND FLUCTUATIONS IN INFLATIONARY UNIVERSE MODELS BASED ON (NEARLY) COLEMAN-WEINBERG GUT MODELS

Paul Joseph Steinhardt
University of Pennsylvania, Philadelphia, Pa. 19104

ABSTRACT

Some recent developments have been made in the study of phase transitions in grand unified theories (GUTs) in which radiatively induced corrections to the effective potential play an important role (e.g. Coleman-Weinberg models). The course of the phase transition is shown to depend sensitively on the choice of parameters of the theory. A restricted class of parameters, excluding the simplest SU(5) GUT model, is found to lead to an inflationary Universe. An heuristic argument is given to show that the spectrum of fluctuations in such models (on observable scales) is scale invariant and a dimensional argument is proposed to explain the amplitude of the resultant fluctuations. Only a much more restricted class of GUT models is found to yield an inflationary Universe with fluctuations that are not so big as to be inconsistent with the isotropy and homogeneity in the observable Universe.

INTRODUCTION

During the past year, tremendous advances have been made in developing a successful inflationary Universe model. The Inflationary Universe, as Guth[1] originally proposed, requires a strongly first order phase transition in the early history of the Universe. The spontaneous symmetry breaking (SSB) transition in grand unified theories (GUTs) that occurs when the temperature of the Universe is $T \approx 10^{14}$ GeV, about 10^{-35} sec after the Big Bang, can be easily adjusted to yield a strongly first order phase transition. The problem with this original scenario is that, in general, the Universe remains trapped in an exponentially expanding (inflating) metastable phase after the Universe supercools below the critical temperature of the phase transition.[2] Bubbles of stable phase are occasionally nucleated, but the nucleation and growth rate is slow compared to the exponential expansion of those regions of the Universe that remain in the metastable phase; as a result, the phase transition is never completed.[2,3] The result is a highly inhomogeneous Universe that is nothing like our own.

About a year ago, attention began to be focussed on phase transitions in GUT models in which the bare mass of the scalar field responsible for the SSB is fine-tuned to be very small compared to the fundamental GUT scale.[4,5] For such theories, radiatively induced corrections to the effective potential play a critical role in determining the shape of the

0094-243X/84/1160147-19 $3.00

effective potential as a function of the Higgs field expectation value, as was first pointed out by S. Coleman and E. Weinberg.[6] We will refer to such models as being <u>nearly</u> Coleman-Weinberg (NCW) models, to distinguish them from the special case of exact Coleman-Weinberg models in which the mass of the scalar field is tuned to be exactly zero.

The NCW models were shown to yield phase transitions in which inflationary expansion occurs but the phase transition is completed. The key feature is that the important inflationary expansion during the phase transition occurs not while the Universe is in the metastable phase, but while it slowly but inevitably evolves towards the stable phase. The resulting cosmology has come to be known as the New Inflationary Universe. In Ref. 7, the New Inflationary Universe model is reviewed along with many subtleties and nuances of the scenario; it is shown how the resultant cosmology resolves the horizon, flatness/oldness, entropy, baryon symmetry, monopole and domain wall puzzles of the standard Hot Big Bang model. This paper should be thought of as a companion to Ref. 7 (i.e. as little review as possible has been given of issues discussed in Ref. 7) in which some more recent progress in the understanding of the NCW transition will be given. The analysis demonstrates the surprising variety of modalities by which the phase transition in these models can proceed (as a function of the parameters) and exhibits where the Inflationary Universe model fits in among these modalities. The generation of density fluctuations in the Inflationary Universe models based on NCW models is discussed and an heuristic argument is given as to why the fluctuation spectrum should be the scale invariant spectrum proposed by Zel'dovich to explain the origin of galaxies. A dimensional argument will be posed for estimating the amplitude of the fluctuations. In order for the amplitude not to be so big as to be inconsistent with the observation of large scale homogeneity and isotropy in the Universe, the parameters of the NCW models have to be excessively fine-tuned. The prospects for the future of the Inflationary Universe model are discussed in the conclusion.

PHASE TRANSITIONS IN NEARLY COLEMAN-WEINBERG MODELS

The purpose of this section is rather technical--to discuss in some detail the nature of the phase transition in NCW models as a function of the parameters of the theory. It will be presumed that sufficient motivation for this discussion has been provided in the introduction and in Ref. 7. The discussion will concentrate on the modes in which the phase transition can take place, only afterwards relating this analysis to inflationary cosmology.

The NCW models are defined by an effective, finite temperature, one-loop scalar potential of the form:

$$V(\phi)=\frac{m^2}{2}\,\phi^2 + B\phi^4\left(\ln\frac{\phi^2}{\sigma^2}-\frac{1}{2}\right) + V_T(\phi) + \frac{1}{2}\,B\sigma^4 \qquad (1a)$$

where

$$V_T(\phi) = \frac{18T^4}{\pi^2} \int_0^\infty dx\ x^2 \ln\{1-\exp[-(x^2+25g^2\phi^2/8T^2)^{1/2}]\} \qquad (1b)$$

represents the one-loop finite temperature corrections. The parameters for the simplest SU(5) GUT model are: $B = 5625/1024\ \pi^2$, which is dimensionless; $\underline{g}$ is the gauge coupling constant; $\sigma = 4.2 \times 10^{14}$ GeV is the only dimensional parameter in the T=0 potential and appears only logarithmically in the potential; and m is a parameter with the dimensions of mass. The scalar field in the simplest SU(5) GUT model is a 24 adjoint Higgs field, Φ, which we have replaced by

$$\Phi = \phi\ \mathrm{diag}\ (1,1,1,-3/2,-3/2)$$

to reduce the scalar field to a single parameter. The coupling constant, g, is fixed so that $g^2/4\pi = 1/40$ at the GUT scale, $\phi = \sigma$, and it obeys a renormalization group equation as a function of temperature given by[8]:

$$\frac{g^2(T)}{4\pi} = 1/2.12\ \ln(T/\Lambda),\ \Lambda=2.74 \times 10^6\ \text{GeV}. \qquad (2)$$

For the analysis of the inflationary phase transition, the mass parameter, m, is not the same as the mass parameter for the theory at zero temperature in a flat space-time. First of all, the theory must be coupled to gravity and this means that there should be an additional term in the effective potential of the form, $\xi R\phi^2$, where R represents the curvature ($\sim\sigma^4/M_p^2$ in the inflationary epoch where $M_p = 1.2 \times 10^{19}$ GeV is the Planck mass).[9] Secondly, Eq. (1) will be used to represent the effective potential during the inflationary phase of the transition during which the Universe approaches a nearly de Sitter vacuum: de Sitter vacuums are characterized by observer-dependent fluctuations with a characteristic temperature of $T_H \approx \sigma^2/M_p$ (Hawking temperature)[10] which induce an additional term in the effective potential of the form $\sim T_H^2\phi^2$. Both of these contributions are proportional to the square of the scalar field and change the effective mass of the scalar field during the inflationary phase. In Eq. (1) it is intended that all these corrections (which remain even as the physical temperature approaches zero) are lumped in the single parameter m, which can be thought of as representing the renormalized de Sitter scalar field mass. Note that this mass may be substantially different from the scalar field mass when the inflationary Universe ends and the ordinary Friedmann-Robertson-Walker (FRW) expansion commences.

The finite temperature effective potential is shown as a function of the scalar field, which acts as the order parameter for the transition, for a range of temperatures in Fig. 1. At high temperatures the effective potential has a single global minimum corresponding to the SU(5) symmetric phase, $\phi = 0$. As the Universe expands and the temperature lowers, the effective potential develops a second local minimum

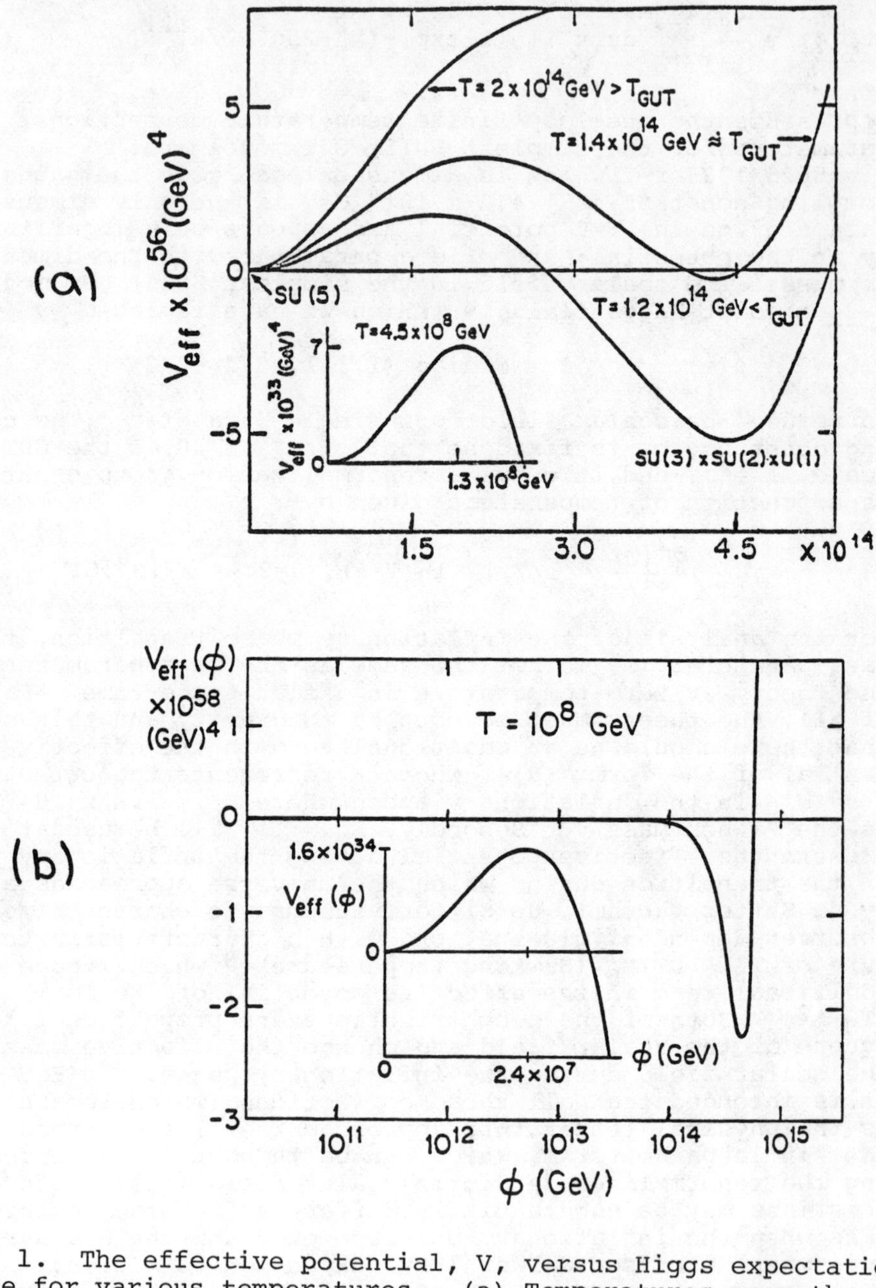

Fig. 1. The effective potential, V, versus Higgs expectation value for various temperatures. (a) Temperatures near the critical temperature. (b) Temperatures below the Hawking temperature and near the metastability limit (see Ref. 7). (For these figures, $g^2/4\pi = 1/45$ at the GUT scale.)

corresponding to a SU(3) x SU(2) x U(1) SSB phase. At a temperature known as the critical temperature $T_c \sim 10^{14}$ GeV, the free energy density of both phases become equal, but the energy barrier prevents the Universe from entering the SSB phase. Even as the temperature drops below the critical temperature, the energy barrier keeps the Universe trapped in what is now a metastable phase and the Universe begins to supercool. For the NCW potentials, the supercooling continues until the thermal contributions to the effective potential become negligible, $V_T(\phi) = 0$ in Eq. (1).

The zero of the potential energy density has been set to correspond to the zero temperature SSB phase because the vacuum energy density of the Universe today - the cosmological constant - is very small (10^{-48} GeV^4) compared to a typical GUT scale energy density (10^{60} GeV^4). The setting of the zero is a fine-tuning based on experimental observation and is not yet explained in any cosmological model. Nevertheless, this means that even after supercooling to near zero (physical) temperature in the SU(5) phase, there remains a tremendous vacuum energy density ($\equiv \varepsilon \sim 10^{60}$ GeV^4).

Einstein's equations for a Robertson-Walker metric include

$$\left(\frac{\dot{R}(t)}{R(0)}\right)^2 = \frac{8\pi}{3M_p{}^2}\rho - \frac{k}{R^2} \qquad (3)$$

where R(t) is the Robertson-Walker scale parameter, ρ is the energy density, and k = +1, 0, -1 in a closed, flat, or open Universe. If the energy density of the Universe approaches a constant vacuum energy density after supercooling, the r.h.s. of Eq. (3) rapidly approaches a constant, $H^2 \sim 8\pi B\sigma^4/6M_p{}^2$. The solution to Eq. (3) is $R(t) = R(0)\exp(Ht)$, which represents the tremendous exponential expansion that characterizes the inflationary phase.

The goal is to arrange a potential where: (1) there is sufficient exponential expansion to solve the cosmological horizon, flatness/oldness, monopole, domain wall and entropy problems:[1,4,5]

$$R(t_e)/R(0) = \exp(60)$$

where the inflation begins at t = 0 and ends at $t = t_e$, and (2) the phase transition to an FRW Universe with SSB phase is completed in a spatial region large enough to contain our observable Universe (radius after the transition ≈ 1 cm) beginning from a region (before exponential expansion) smaller than a single causal horizon volume (radius $\approx 10^{-25}$cm). Again we refer the reader to Ref. 7 for further justification for these statements.

Although the Universe supercools to physical temperature T = 0, an observer in the inflationary phase still measures an effective temperature, as was mentioned before. However, that temperature, in fact all features of the inflationary (de Sitter) phase, remain very nearly invariant with time even as the Universe expands. The potential that describes this phase is Eq. (1) with $V_T(\phi) = 0$, where all parameters are meant to include the effects of all de Sitter fluctuations, renormalizations and gravitational couplings. Recently, A. Albrecht, L. Jensen and I studied in detail the different modes in which the transition from the de Sitter inflationary phase to the SSB phase can occur as a function of the parameters, and I would like to give a brief overview of our results.[11]

The mode of the phase transition is determined by two parameters: the mass parameter (m^2) and the dimensionless coupling constant, B. It is convenient to rewrite the effective potential as

$$V(\phi) = -\frac{1}{4}\lambda\ (\phi)\phi^4 \tag{4a}$$

where

$$\lambda(\phi) = 4B(\frac{1}{2} - \ln \frac{\phi^2}{\sigma^2}) \tag{4b}$$

is the effective dimensionless coupling for ϕ. Both parameters will be evaluated in the de Sitter phase at a typical energy scale O(H). We will study the phase transition as a function of $m^2(H)$ and $\lambda(H)$.

CASE I: $m^2 > 0$.

For this case there remains a barrier even as the physical temperature approaches zero. Normally, one would suppose that the transition occurs through nucleation of bubbles of stable phase that grow, converting metastable phase to stable phase, as first outlined by Coleman and Callan.[12] However, gravitational effects become very important as the barrier becomes small and a careful analysis must be made.

Coleman and De Luccia[13] were the first to consider the gravitational effects on the bubble nucleation process. The effects, they concluded, are very small: closer examination[11] shows that gravity tends to thicken the walls of the bubble (both the interior and exterior radius). The "thickening" is negligible unless $m \lesssim \sqrt{15}\ H$ (in the units we have chosen). A value of $m = \sqrt{15}\ H$ corresponds to the potential in which the curvature, R, is coupled to a massless scalar field in a conformally invariant way[9] (and de Sitter fluctuation effects have been ignored). To lowest order in (H/σ), the action of the bubble is given by

$$A_{CD} = \frac{15}{2}^2\ \frac{8\pi^2}{3}\ \frac{1}{\lambda} \tag{5}$$

For $m \lesssim \sqrt{15}\ H$, Hawking and Moss[14] have argued that the Coleman-De Luccia bubble solution does not exist. They show, using a thin-wall approximation, that bubbles would become larger than the de Sitter event horizon for $m \lesssim \sqrt{15}\ H$; such a bubble would be inconsistent with the fact that coherent fluctuations cannot be created on scales greater than the de Sitter horizon. Their published arguments are somewhat suspect, however, since the thin-wall approximation is entirely untrustworthy in the limit of a small (NCW) barrier; in Ref. 11 we have proven this result in a more general way. Hawking and Moss noted that another semi-classical solution becomes relevant for $m \leq \sqrt{15}\,H$ - one in which $\phi = \phi_{top}$ everywhere, where $V(\phi_{top})$ is the absolute maximum of $V(\phi)$ (see Fig. 2). The solution can be interpreted as a bubble solution with an infinitely thick wall ($\phi=\phi_{top}$ defines the wall of a Coleman-De Luccia bubble solution) and an infinitesimal interior. The action of the solution can be shown to be[11]

$$A_{HM} = \frac{2\pi^2}{3} \frac{1}{\lambda} \frac{m^4}{H^4} \tag{6}$$

Comparing this result to Eq. (5) one finds that this solution obtains the lower action for $m \leq \sqrt{15}\,H$, and becomes the dominant Euclidean semi-classical contribution to the decay of the false vacuum. Thus, for $m \geq \sqrt{15}\,H$ the transition occurs by the Coleman-DeLuccia bubble nucleation mode and for $m \leq \sqrt{15}\,H$ the transition occurs by the Hawking-Moss mode.

The Hawking-Moss solution has been interpreted[15] by some to mean that the Universe "jumps out of the metastable phase" all at once - coherently - over all of de Sitter space. If true, the result is quite remarkable because it implies that an event can occur coherently over scales very large compared to the de Sitter event horizon! I would like to offer a criticism of this interpretation. The interpretation of a semi-classical tunneling solution can only be made by studying conditions long after the tunneling event; the semi-classical solution cannot be trusted to give information about the tunneling process itself. The Coleman-DeLuccia solution, for example, is interpreted as a bubble nucleation process only because long after the tunneling the solution describes a (nearly) classical expanding bubble. By extrapolating that expansion backwards in time one can surmise that the bubble "nucleated" at some approximate time, but details of the nucleation process itself (e.g., how did the bubble appear when it first nucleated) cannot be reliably inferred from the semi-classical solution. The problem with interpreting the Hawking Moss solution is that, because $\phi = \phi_{top}$ is a position of unstable equilibrium, the future of the tunneling state cannot be inferred without more information than is included in the semi-classical solution itself.

What is probably the case is that quantum fluctuations drive the Universe away from $\phi = \phi_{top}$, either towards the metastable phase or towards the stable phase. A typical fluctuation will drive a region of radius of order the de Sitter Hubble radius ($O(H^{-1})$) coherently towards one phase or the other. Regions that return to the metastable phase exponentially expand (because the vacuum energy density is large in those regions) and rapidly dominate the volume of the Universe. The regions that are driven towards the stable

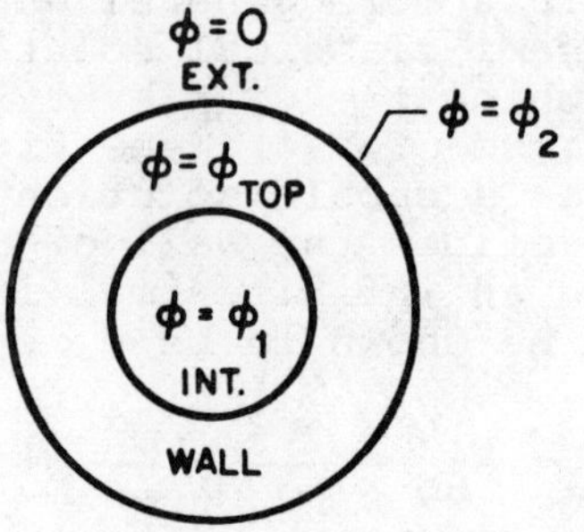

(a)

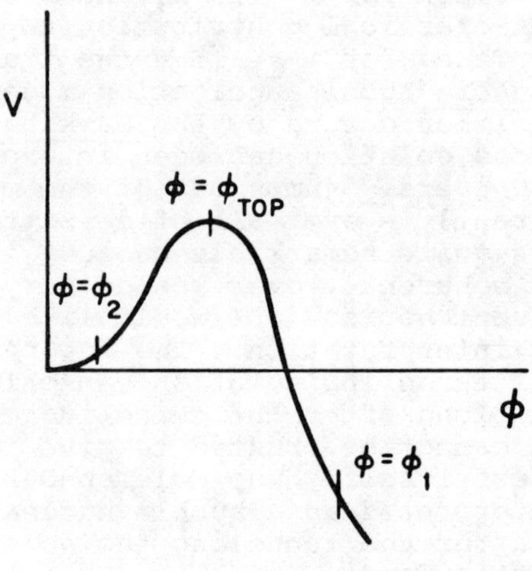

(b)

Fig. 2. The effective potential vs. ϕ for a small but positive m^2. For a Coleman-DeLuccia solution, ϕ varies from ϕ_1 in the interior of the bubble (with radius $\lesssim H^{-1}$) to ϕ_2 on the exterior of the bubble. Regions where $\phi \approx \phi_{top}$ represent the wall of the bubble. For the Hawking Moss solution $\phi = \phi_{top}$ everywhere!

phase will eventually reach that stable phase and stop exponentially expanding. If the potential on the stable phase side of the barrier is steep (so that there is no significant inflation during the evolution towards the stable phase) these regions will enter the stable phase with a typical volume $O(H^{-3})$. They will then begin to grow and convert the Universe from metastable to stable phase, in essentially the same way as Coleman-DeLuccia bubbles do. Far in the future the Universe appears to contain many (small) islands of stable phase with edges expanding into the metastable phase, but most of the Universe lies in exponentially expanding metastable phase. If this interpretation is true, then the asymptotic state reached from the Hawking Moss solution appears to be very little different from ordinary bubble nucleation, although the stable regions created by the fluctuations may be much more irregular than spherical Coleman-DeLuccia bubbles. Certainly, it would be inaccurate to describe this process as the Universe jumping out of the metastable phase all at once!

Nevertheless, we will treat the Coleman-DeLuccia and Hawking-Moss semi-classical solutions as representing two different modes in which the phase transition can occur. In Fig. 3 is shown a plot of the regions of parameter space in which two modes are relevant. (Note: Only the regions above the abscissa correspond to $m^2 > 0$.) In either case, the phase transition is never completed due to the exponential expansion of the metastable phase - the same problem that plagued Guth's original scenario.[2] If the potential is very flat on the stable phase side of the barrier, regions (or bubbles) of stable phase may exponentially expand to a size much larger than our observable Universe before entering the stable, FRW phase. In this case, our own Universe would lie inside just such a region. In fact, the New Inflationary Universe based upon NCW transitions has often been described in just such terms. However, when $m^2 \gtrsim 0$ becomes very small, one finds that the semi-classical approximation breaks down altogether. The signal for this is that the Euclidean action becomes of order unity (in dimensionless units) and perturbations on the semi-classical solution become very important. In fact, as was argued in Ref. 5, the barrier becomes essentially negligible and the metastable phase becomes unstable. (A few bubbles may nucleate before the physical temperature reaches zero, but these are rare and negligible for this discussion.) At this point, quantum fluctuations drive different regions of the Universe towards one SSB or another, and each "fluctuation region" in the Universe evolves towards a stable state. In this picture, the Universe breaks up into domains, each of which approaches one of a continuum of stable phases and no part of the Universe is left in the de Sitter phase. As in Ref. 7 we term this third phase transition mode the "spinodal decomposition of the Universe" from the term for the analogous process in condensed matter physics.

Where does inflation fit among the Coleman-DeLuccia, Hawking-Moss, and spinodal modes? Sufficient inflation depends sensitively on the initial spectation value of ϕ, $\phi(0)$,

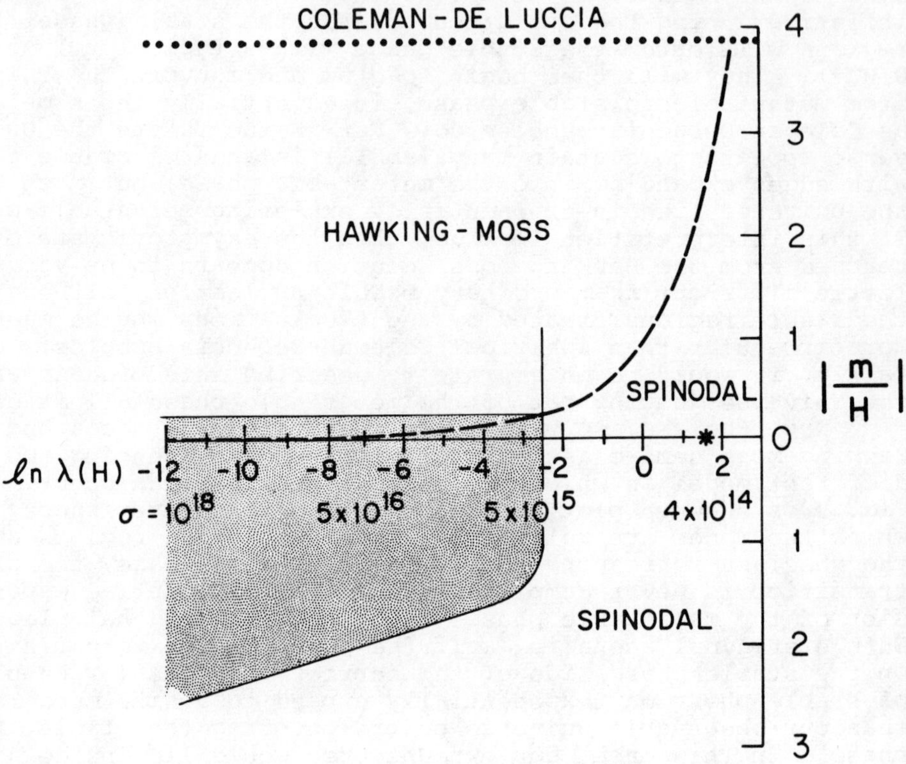

Fig. 3. Phase transition modes as a function of the parameters of a (nearly) Coleman-Weinberg Theory. The abscissa shows the effective dimensionless coupling of the de Sitter phase, $\lambda(H)$. On the same axis is shown the value of σ necessary to maintain the same Hubble parameter as in the simplest SU(5) GUT model (~ 10^9 GeV). The axis represents the mass parameter, $|m|$. The upper part of the axis is for $m^2 > 0$, the lower axis is for $m^2 < 0$. Above the dotted line the Coleman-DeLuccia mode dominates. Between the dotted and dashed lines the Hawking Moss mode dominates and below the dashed line the spinodal mode dominates. The hatched region is the region of parameter space for which inflation can be obtained. Only for $\lambda(H)$ less than 10^{-11}, though, are the associated fluctuations small enough to be consistent with observation.

in a bubble or fluctuation that takes a region of the Universe to the stable phase side of the barrier. The initial value of ϕ grows, to first approximation, according to the semi-classical equation:

$$\ddot{\phi} + 3\,\frac{\dot{R}}{R}\dot{\phi} = -\,V'(\phi) + \alpha H^4 \tag{7}$$

where α is constant. The constant term on the r.h.s. of the equation approximates the effects of quantum de Sitter fluctuations that Vilenkin and Ford[16] and Linde[17] showed are non-negligible in NCW potentials when ϕ is very small.[27] Using Eq. (7) we have determined for what choices of parameters the evolution of ϕ away from $\phi(0)$ is slow enough that sufficient inflation can be achieved in Eq. (3): this represents the hatched region of Fig. 3. As can easily be surmised, the Coleman-DeLuccia mode is irrelevant for inflation and only a very small region of parameter space associated with the Hawking-Moss solution is relevant. The spinodal decomposition mode is the best approximation to the inflationary Universe phase transition in $m^2>0$ NCW models.

Indicated by the star in Fig. 3 is the value of $\lambda(H)$ for the usual choice of SU(5) GUT parameters. The fact that this lies outside the region of parameter space relevant to inflation is due primarily to the Vilenkin-Ford-Linde[16,17] effect represented by the constant term in Eq. (7). If this were ignored, $\lambda(H)$ would have been just outside the resulting hatched region of the figure.[11] (Note that $g^2/4\pi$ was set to 1/40 at the GUT scale, instead of 1/45 that was used in some of the previous references[5]; the larger is the value of g^2, the less flat is the potential and the harder it is to achieve inflation.)

CASE II: $m^2 < 0$.

If $m^2 < 0$, the barrier between the metastable and stable phases disappears altogether before the temperature reaches zero. The metastable symmetric phase becomes unstable and thermal as well as quantum fluctuation can act to drive the Universe away from the unstable symmetric phase. For the cases where inflation is possible, the barrier disappears when the physical temperature, T, is much less than the Hawking temperature of the de Sitter phase, and so thermal effects can be ignored. The phase transition proceeds according to the spinodal decomposition mode discussed previously. If a typical fluctuation drives the expectation value of ϕ to a value of order the Hawking temperature at the time the barrier disappears, the degree of subsequent inflation of a fluctuation region can be approximated from Eqs. (3) and (7). Sufficient inflation of a fluctuation region is represented by the hatched region below the abscissa in Fig. 3. The GUT value of $\lambda(H)$ lies outside the hatched region.

Our conclusion is that inflation occurs among the NCW

SU(5) GUT models almost exclusively through the spinodal decomposition mode. The mass parameter must be tuned to be of order H^2 or less, representing a fine-tuning of m^2 by 10 orders of magnitude compared to the typical GUT scale. The quantum fluctuation effects computed by Vilenkin and Ford[16] and Linde[17] are sufficient to drive the expectation value of ϕ too fast down the potential to achieve significant inflation for the usual choice of $\lambda(H)$ in the simplest SU(5) GUT model. We will see in the next section that the analysis of the fluctuations in the Inflationary Universe will lead to even more stringent conditions necessary for NCW models to yield viable Inflationary Universe scenarios.

FLUCTUATIONS IN THE INFLATIONARY UNIVERSE

The Inflationary Universe model was designed to solve the cosmological horizon, flatness, monopole, domain wall, entropy, and baryon asymmetry problems within a simple and efficient model.[7] Although the usual choice of parameters in the simplest SU(5) NCW model seems to be out of range to yield an Inflationary Universe, there is much hope that a yet newer version of the Inflationary Universe is possible.[18-20] During the course of the past summer, yet another potential success of the Inflationary Universe model was discovered: the Inflationary Universe predicts primordial fluctuations in the energy density of the Universe with just the qualitative spectrum that many cosmologists believe is necessary to explain the origin of galaxies!

For many years cosmologists have debated the issue of what spectrum of fluctuations is necessary to account for fluctuations. Until the inflationary Universe cosmology, the fluctuation spectrum could not be derived from first principles because the spectrum is dependent upon the initial fluctuations that existed when the Universe first emerged from the big bang. To make matters worse, in order to form galaxies, coherent fluctuations have to exist on distance scales large compared to the Hubble radius (causal horizon radius) in the Hot Big Bang Model. Nevertheless, some cosmologists contented themselves with working backwards - from observational data on the organization of galaxies in the Universe, they hypothesized what kind of fluctuation spectrum is necessary to explain it without determining the source of those fluctuations.[21] Of course, there is much debate about the interpretation of the observational data and whether or not a given fluctuation spectrum can really lead to a Universe in agreement with the observational data; in my opinion, the issue is still very far from being resolved. Nevertheless a large fraction of cosmologists have come to believe that a "scale-invariant" spectrum of adiabatic fluctuations, as first proposed by Zel'dovich[22], is what is required.

The fluctuation spectrum is characterized by the amplitude of the fractional energy density fluctuation in the energy density $\delta\rho/\rho$, as a function of wavelength, ℓ. An adiabatic fluctuation is one in which the fluctuation in the temperature and in the number of baryons is the same, as is expected in any GUT. For wavelengths that are large compared to the Hubble radius, which sets the scale of the causal horizon, the fluctuation amplitude, $\delta\rho/\rho$, is not a coordinate gauge invariant quantity - its numerical value depends upon how one fixes the hypersurfaces of constant time. However, for wavelength scales less than the Hubble radius, the fluctuation amplitude represents a physical (gauge invariant) quantity; for example, in a radiation dominated Universe $\delta\rho/\rho$ represents the amplitude of an acoustic wave. For this reason, $\delta\rho/\rho$ is only measured on scales less than or equal to the Hubble horizon. Cosmologists, in fact, define the spectrum of fluctuations by plotting $\delta\rho/\rho$ when the wavelength is just equal to the Hubble radius; this quantity we call $\delta\rho/\rho|_H$. In a FRW Universe, the Hubble radius grows with time, so a plot of $\delta\rho/\rho|_H$ versus ℓ measures the fluctuations on different scales at different times (as they fall within the expanding Hubble radius). In this way, the amplitude, $\delta\rho/\rho|_H$, represents the "virgin perturbation amplitude" before causal physics has had a chance to act coherently on the fluctuation and alter it.

Every physical scale, ℓ, at any given time in the Universe can be characterized by a unique "comoving" scale, in which the expansion of the Universe has been divided out, $\ell = R(t)\,\ell_{comoving}$. On the comoving scale associated with galaxies, we require that $\delta\rho/\rho|_H$ be greater than 10^{-4} in order to have a big enough fluctuation to eventually condense and form galaxies. However, the fluctuation on the largest observable comoving scales cannot be greater than 10^{-4} or quadrupole anisotropies would have already been observed in the microwave background. Also, if $\delta\rho/\rho|_H$ is greater than unity on any observable scale, many black holes should form on that scale, too many to be consistent with observations. The simplest spectrum that is consistent with all these constraints is one in which $\delta\rho/\rho|_H$ is nearly constant ($\sim 10^{-4}$) over all scales and this is just the spectrum proposed by Zel'dovich.[22]

Zel'dovich's motivation for proposing this simple spectrum was not based solely on this simplistic argument. He hoped to explain the organization and origin of galaxies from such a spectrum. Basically, he argued that, beginning from such a spectrum, fluctuations on small scales would be damped out by various dissipative mechanisms. The first surviving fluctuations would be on the galaxy cluster or supercluster scale and it is on this scale that condensations would first be formed. Zel'dovich then argued that such fluctuations would condense in a hydrodynamically unstable mode, breaking up into subsections that collapse into flat, "pancake" surfaces from which galaxies form. One implication of such a "pancake theory" is that there should be large observable voids separating the pancakes, and already preliminary observation of such voids have been reported in the popular press. (Note: Voids are required in this picture, but may not be unique to the scale-invariant spectrum.) Also, the nature of the "dark matter" in the Universe which accounts for most of the energy density is an important consideration, whether it be massive neutrinos, gravitinos or baryons. If, as the inflationary Universe predicts, Ω is very nearly 1,[1] a massive neutrino species (or another light relic species) which dominates the mass density is necessary or else deuterium is unproduced and ^{4}He is overproduced during primordial nucleosynthesis.[23] In fact, only in a Universe dominated by a light relic species can adiabatic perturbations with size $\delta\rho/\rho = 10^{-4}$ (not too large to be inconsistent with the isotropy of the microwave background) grow sufficiently to produce galaxies by the present epoch.

Thus, it is truly remarkable that during the course of the last summer four independent groups using five independent methods studied the fluctuations that should be produced in an inflationary Universe and concluded that the spectrum should be an essentially scale invariant spectrum[24-27] just as Zel'dovich ordered. Furthermore, the amplitude of the spectrum was found to be calculable, and dependent upon the details of the GUT theory responsible for the inflation. The bad news, as we will see, is that the amplitude of the fluctuations for the simplest SU(5) Coleman-Weinberg GUT model is much too big, but we will try to argue in the concluding section that this is only a temporary setback. The result is especially remarkable because it was not anticipated and it was discovered as an automatic feature of a model that was already tremendously successful in solving cosmological conundrums. Furthermore, it should be re-emphasized that this is the only cosmological model that has any hope of explaining our present Universe, and for which the spectrum of fluctuations can be derived from first principles.

The derivations of the four independent approaches are quite complicated and it is very difficult to see how one derivation is related to another. This is due to the fact that the different derivations are based on different choices for the hypersurfaces of constant time. The "history" of a fluctuation on a given scale appears different for different hypersurfaces, but, as mentioned above, by the time the fluctuation re-enter the Hubble horizon in the FRW phase, the amplitude is the same independent of the choice of constant time hypersurfaces. Although it is very difficult to describe any single one of the derivations, I will attempt to provide a heuristic argument as to why the qualitative spectrum is found to be scale-invariant and a dimensional argument as to why, in the simplest SU(5) Coleman-Weinberg model, the quantitative value found for the amplitude is too large.

<u>Why is the fluctuation spectrum in the inflationary Universe scale invariant?</u> The Hubble radius, H^{-1}, represents an important scale in the analysis of the creation and the evolution of energy density perturbations. During the de Sitter or inflationary phase, H (and other parameters) are roughly

constant, $H^2 \simeq 8\pi V(0)/3M_p^2 \simeq M_G^4/M_p^2$. The scale factor R(t) grows exponentially with time during the de Sitter phase, undergoing one e-folding in a time interval $0(H^{-1})$. Microphysics can only operate coherently on physical scales less than $0(H^{-1})$ during the de Sitter phase.

The basis of the inflationary Universe is that all scales we observe in our Universe today have a physical wavelength, ℓ, that is much less than the Hubble radius when the inflationary epoch begins; the horizon problem, for example, is solved in the inflationary scenario because our observable Universe lies within a causal horizon volume (with radius $0(H^{-1})$) before the inflationary epoch begins. Microphysics (quantum fluctuations, etc.) can create and alter tha amplitude of perturbations on scales for which $H\ell \ll 1$. As inflation proceeds, the physical scale of a perturbation grows until $H\ell \approx 1$, and then onward until $H\ell \gg 1$. Once the physical scale of the perturbation has grown such that $H\ell \approx 1$, microphysics cannot act coherently on that scale and alter the amplitude of the physical perturbation: the perturbation "freezes out" at whatever amplitude it had at the time $H\ell \approx 1$. Since the inflationary (de Sitter) epoch is essentially time translationally invariant, the same microphysics determines the amplitude of the perturbation on different scales as they expand beyond the Hubble horizon ($H\ell \approx 1$) at different times. Once the Universe reheats and re-enters the FRW phase, the Hubble radius increases (at the rate characteristic of an FRW Universe); $H\ell$ begins to shrink until $H\ell \approx 1$ once again (now in the FRW phase) and the scale of the perturbation is said to "re-enter the horizon". Since no microphysics could change the perturbation when $H\ell \gg 1$, and since the perturbation on each scale "freezes out" at the same amplitude, each scale "re-enters the horizon" with the same amplitude of perturbation, $\delta\rho/\rho|_H$. Thus, the spectrum is scale invariant.

<u>What is the amplitude of the fluctuation spectrum?</u> The methods used in Refs. 24-27 not only show that the fluctuation spectrum is scale invariant, but determine the amplitude of that spectrum. In fact, the methods used in Ref. 27 show explicitly that the amplitude is very insensitive to the equation of state during reheating and through the FRW epoch. I do not believe that there is any simple argument in general to explain the result, but the expression for the fluctuations amplitude is found to be:

$$\delta\rho/\rho\Big|_H = (4 \text{ or } \tfrac{2}{5})\, H\Delta\phi/\dot{\phi}\Big|_{t_{freeze}} \qquad (8)$$

$\dot{\phi}(t)$ is the time variation of the scalar field whose evolution signals the SSB phase and is responsible for the inflationary phase; $\Delta\phi$ is the fluctuation in ϕ, and the prefactor is 4 (or 2/5) if the Universe is radiation-dominated (or matter-dominated) when a given scale re-enters the Hubble horizon in the FRW phase. For comparing to the anisotropies in the microwave background, the 2/5 prefactor is appropriate. Notice that the answer only depends upon the values of parameters at a time t_{freeze} when the given scale of interest expands beyond the Hubble radius during the inflationary phase, as seems sensible from our heuristic argument.

It is also clear from Eq. (8) that the spectrum is not precisely scale-invariant; during inflation $\Delta\phi$ is essentially a constant, H^2, as determined by the quantum fluctuations of the scalar field (in a de Sitter vacuum, there is a characteristic temperature, $H/2\pi$, the Hawking temperature).[10] The denominator, $\dot{\phi}$, is the classical time variation of the scalar field as determined by Eq. (7). During inflation, ϕ and $\dot{\phi}$ increase as the field rolls towards the SSB phase and have somewhat different values when perturbations on different scales freeze out as they expand beyond the de Sitter Hubble radius. The result is that the fluctuation spectrum has scale dependence that is logarithmic (i.e. Eq. (8) is in fact scale invariant up to logarithmic factors. The freeze-out time depends only logarithmically on scale because the scale parameter grows exponentially with time.) The actual evaluation of Eq. (8) is found to be if (2/5 is used in Eq. (8)):

$$\delta\rho/\rho\big|_H \approx 10 \quad \lambda^{\frac{1}{2}} \, [1 + \ln(10^{28}\mathrm{cm}/\ell)]^{3/2} \qquad (9)$$

The dimensionless parameter, $\lambda(H)$, is 10 in the simplest SU(5) GUT model and so $\delta\rho/\rho|_H$ is of order 10 on the largest scales we can observe today ($\ell = 10^{28}$cm). The result is roughly six orders of magnitude too big and much too big to be consistent with the isotropy of the microwave background. It might be pointed out that we have found $\delta\rho/\rho|_H$ is greater than unity, so it is possible, in principle, that non-linear effects can save us; this is probably wishful thinking.)

Although it is difficult to derive Eq. (8) explicitly, it is not difficult to explain why, on dimensional grounds, the answer we have obtained is of order unity. The key is that during inflation ϕ lies very close to the top of the potential where the potential is essentially flat. The GUT scale, ϕ, only appears logarithmically in the NCW potential and the only dimensionful parameter that determines the evolution of ϕ (see Eq. 7) and the perturbations in ϕ (Hawking temperature $\sim H/2\pi$) is the Hubble parameter, H. Once ϕ grows much beyond H its evolution is very rapid and the Universe begins to reheat. The fractional perturbation, $\delta\rho/\rho|_H$, we have argued, depends only upon the properties of the inflationary epoch. Because it is dimensionless and there is only one relevant dimensionful parameter, $\delta\rho/\rho|_H$ can only be a function of dimensionless quantities, such as $\lambda(H)$. The dimensionless coupling constant, $\lambda(H)$, for the simplest SU(5) model is of order unity, so it is difficult to imagine obtaining the desired value of $\delta\rho/\rho|_H = 10^{-4}$ that we need for the fluctuation spectrum to be consistent with the isotropy of the microwave background (and, in fact, our calculations seem to show that all dimensionless factors tend to conspire to make the amplitude larger, rather than smaller.)

To obtain the desired value, $\delta\rho/\rho|_H = 10^{-4}$, the dimensionless coupling, $\lambda(H)$, must be reduced by over ten orders of magnitude from its value in the simplest GUT mode. By simply reducing $\lambda(H)$ and not other parameters, though, $H \propto \lambda$ is also reduced and then inflation is reduced. To keep H fixed while reducing $\lambda(H)$, the GUT scale, σ, has to be increased. Along the abscissa in Fig. 3 is shown not only the value of $\lambda(H)$ but the associated value of σ necessary to keep H fixed. The severe constraint coming from the fluctuation analysis forces parameters to lie in a very small range where $\sigma = 10^{18}$ or 10^{19} GeV in the GUT model. By tuning $\lambda(H)$ to be small and σ to be large, the potential is designed to greatly increase the flat "rollover" region of the potential. As a result, $\dot{\phi}$ is significantly greater than H^2 towards the end of the inflationary epoch. Since $\dot{\phi}$ is greater than H^2 when observable scales in our Universe expand beyond the Hubble radius in the inflationary epoch, a smaller value of $\delta\rho/\rho|_H$ is obtained in Eq. (8) for such scales. One might have guessed from the dimensional argument alone that the only hope of reducing the dimensionless quantity, $\delta\rho/\rho|_H$, in a theory with only one dimensionful parameter is to introduce abnormally small dimensionless parameters (in this case, $\lambda(H)$). Since there is no GUT model for which we expect parameters to lie in the desired range, the New Inflationary Universe model based on (nearly) Coleman-Weinberg potentials is dead.

FUTURE PROSPECTS FOR THE INFLATIONARY UNIVERSE

In spite of all the criticisms of the New Inflationary Universe model that have been discussed in this brief review, inflationary cosmology is very sound, even thriving. The key notion that was garnered from the studies of the Coleman-Weinberg model - the idea of a slow "rollover" transition in which inflation occurs in evolving stable phase rather than in trapped metastable phase - can be carried over to a broader class of models. This notion of the slow transition is crucial for two reasons. First it insures that the transition can be completed because the inflating Universe evolves slowly but inevitably towards the stable phase. Second, unlike Guth's original picture where it was hoped that the transition would be completed by the rather discontinuous and inhomogeneous process of bubble nucleation and collision, the slow transition is a continuous process in which all physical quantities can be tracked in a smooth way. This last feature is crucial for analysing the creation and growth of fluctuations, for example.

Already there has been much thought given to incorporating the inflationary cosmology within a supersymmetric potential where slow "rollover" potential can occur without fine-tuning of parameters.[18,19,20] Criticism of these approaches will not be given here. What is clear from these analyses is that the fine-tuning of m^2 and $\lambda(H)$ that is required for

NCW models is not a fundamental problem, but can be fixed rather naturally in a supersymmetric model, for example.

No matter which is the ultimate theoretical scheme, that scheme will have to solve the gauge hierarchy problem. Almost every scheme that has been attempted so far to solve the gauge hierarchy problem builds the large hierarchy between the Weinberg-Salam and GUT scales from a much smaller, more natural hierarchy, or regenerates several large ($\gtrsim 10^{12}$ GeV) scales dynamically. Without looking at the details of any one of these approaches it should be seen that this is good news for the inflationary Universe. The problem with the inflationary Universe based on NCW models, we have seen, is that there is only one relevant dimensionful quantity, H. The Hubble parameter is a ratio of a GUT scale to a Planck scale. In a theory in which there is an additional large mass scale, there are many ways of building different dimensionless expressions for $\delta\rho/\rho|_H$. One expects $\delta\rho/\rho|_H$ to be a function of M_1/M_2 to some power, where M_i are the large scales, and the ratio can easily be adjusted to make $\delta\rho/\rho|_H$ 10^{-4} (see Ref. 18, for example). Thus, in my opinion, solutions to the gauge hierarchy problem and the problem of obtaining a viable Inflationary Universe model are likely to be intertwined.

I would like to thank my collaborators, Andy Albrecht, Jim Bardeen, Lars Jensen and Mike Turner, without whom much of the work reported at the conference would not have been completed. I would like to thank the La Jolla Institute and especially Frank Henyey for the kind hospitality offered during the Workshop. Last, but not least, I would like to thank Glennys Farrar and Frank Henyey for arranging such a stimulating conference, as well as outstanding weather. My only regret is that I never really did see a whale.

This work was supported in part by DOE Grant EY-76-C-02-3071 and by a DOE Outstanding Junior Investigator Grant.

References

1. A Guth, Phys. Rev. D23, 347 (1981).
2. A. Guth and E. Weinberg, MIT Preprint CTP-950, to be published in Nuclear Physics.
3. S. W. Hawking, I. G. Moss and I. M. Stewart, Phys. Rev. D26, 2681 (1982).
4. A. D. Linde, Phys. Lett. 108B, 389 (1982).
5. A. Albrecht and P. J. Steinhardt, Phys. Rev. Lett. 48, 1220 (1982). A. Albrecht, P. J. Steinhardt, M. S. Turner, and F. Wilczek, Phys. Rev. Lett. 48, 1437 (1982).
6. S. Coleman and E. Weinberg, Phys. Rev. D7, 788 (1973).
7. P. J. Steinhardt, "Particle Physics and the New Cosmology", Proceedings of the 1982 APS Meeting on Particles and Fields, Baltimore, Md. Penn preprint 0210T (1982).
8. M. Sher, Phys. Rev. D24, 1847 (1981).
9. N. D. Birrell and P.C.W. Davies, Quantum Fields in Curved Space (Cambridge University Press, 1982).

10. G. Gibbons and S. W. Hawking, Phys. Rev. D15, 273 (1977).
11. A. Albrecht, L. Jensen and P. J. Steinhardt, to be published (1983).
12. S. Coleman, Phys. Rev. D15, 2929 (1977).
C. Callan and S. Coleman, Phys. Rev. D16, 1762 (1977).
13. S. Coleman and F. DeLuccia, Phys. Rev. D21, 3305 (1980).
14. S. Hawking and I. Moss, Phys. Lett. 110B, 35 (1982).
15. E. Mottola and A. Lapides, Santa Barbara preprint NSF-ITP-82-101 (1982).
16. A Vilenkin and L. Ford, Tufts Preprint (1982), to be published.
17. A. D. Linde, Lebedev Physical Institute Preprint (1982). See also T. S. Bunch and P.C.W. Davies, Proc. Roy Soc. A360, 117 (1978).
18. P. J. Steinhardt, "Natural Inflation", Penn preprint UPR-0198T (1982). A. Albrecht, S. Dimopoulos, W. Fischler, E Kolb, S. Raby and P. Steinhardt, Proceedings of the Third Marcel Grossmann Meeting on the Recent Developments in General Relativity, Los Alamos Preprint LA-UR 82-2947.
19. J. Ellis, D. V. Nanopoulos, K. Tamvakis, CERN preprint TH.3418.
20. D. V. Nanopoulos, K. A. Olive and M. Srednicki, CERN Preprint TH.3477 (1982).
21. P.J.E. Peebles, The Large Scale Structure of the Universe (Princeton Univ. Press, 1980).
22. Ya. B. Zel'dovich, Mon. Not. R. Astr. Soc. 160, 1P (1972); E. R. Harrison, Phys. Rev. D11, 2726 (1970).
23. J. Bond and A. Szalay, in Neutrino '81, eds. R. J. Cence, E. Ma, A. Roberts, High Energy Physics Group, Univ. of Hawaii (1981).
24. S. W. Hawking, Phys. Lett. 115B, 295 (1982).
25. A. Guth and S.-Y. Pi, Phys. Rev. Lett. 49, 1110 (1982).
26. A. A. Starobinskii, Proceedings of the 1982 Nuffield Workshop on the Very Early Universe (1982).
27. J. M. Bardeen, P. J. Steinhardt, and M. S. Turner, Penn Preprint UPR-2020T (1982).

FINITE FOUR DIMENSIONAL SUPERSYMMETRIC THEORIES

P. West
Mathematics Department, King's College
Strand, London WC2

ABSTRACT

It is shown that there exist two classes of supersymmetric four dimensional field theories which are finite to all orders of perturbation theory. One class consists of N = 2 supersymmetric Yang-Mills coupled to N = 2 super-matter for which the representation content of the matter fields must be chosen so as to give vanishing contribution to the one loop beta-function. An explicit calculation verifying this result to two loops is given. The other class is generated by adding specific combinations of mass terms as well as terms cubic in the scalar and pseudo-scalar fields to the N = 4 supersymmetric Yang-Mills theory.

1. INTRODUCTION

Immediately after the discovery of supersymmetry it was realized that the Wess-Zumino model had remarkable ultraviolet properties.[1] Since then it has become apparent that cancellation of ultraviolet divergences is a general feature of supersymmetric theories.[2] The purpose of this talk is to examine the ultraviolet behavior of theories of rigid extended supersymmetry. The table below gives a list of these theories;[3] the N = 1 supersymmetry theories being included for comparison.

N \ Spin	1	1	2	2	4
Spin 1	-	1	1	-	1
Spin ½	1	1	2	2	4
Spin 0	2	-	2	4	6

The table stops at four supercharges (N = 4) since for N > 4 the irreducible representation of supersymmetry have spins greater than one. The N = 3 theory has the same particle content as the N = 4 theory and is thought to be identical to the N = 4 Yang-Mills theory. The inclusion of spin $\frac{3}{2}$ is known[4] to require a spin 2 in order to avoid acausal propagation, and as is well known, the inclusion of spin two leads to theories with local supersymmetry.

The only two N = 2 rigid theories are N = 2 Yang-Mills[5] and N = 2 matter (the hypermultiplet)[6]. The coupling and properties of these theories will form a central part of this talk.

The supersymmetric N = 4 Yang-Mills theory[7] is the last entry in the table and it represents the most symmetrical consistent theory that we know. Here, the use of the word consistent signifies renormalizability and absence of acausal behavior. This theory is unique up to the choice of the gauge group and the coupling constant. This theory has been shown to be finite up to three loops[8]. In section two we will review the arguments which establish that this theory is finite to all orders.

The main subject of this talk is to show that the N = 4 Yang-Mills theory is not the unique finite theory of extended supersymmetry. In section three we will show that an arbitrary N = 2 theory containing spins of one and less is finite above one loop; the Callan-Symanzik beta function being given exactly by the one loop contribution. We will show that it is possible to arrange the representation content of the N = 2 matter so as to make the one loop beta function vanish. These models then constitute a class of theories which are finite to all orders of perturbation theory.

In section four an explicit calculation varifying the results of section three is given. The N = 2 theory is expressed in terms of N = 1 superfields and the N = 1 super-Feynman rules are used to calculate the two loop propagators of the theory and consequently find the two loop ultraviolet behavior.

In section five another class of finite supersymmetric four-dimensional field theories is given. All terms of dimension three and less which are compatible with Lorentz and gauge invariance, but respecting no other symmetries are added to the N = 4 supersymmetric Yang-Mills theory. The necessary conditions for finiteness are derived and it is found that, provided certain relations between the coefficients of these additional terms hold, the theory maintains its finiteness properties. One of these relations is that the supertrace of the square of the added masses is zero.

2. FINITENESS OF N = 4 YANG-MILLS

There are three known arguments for the finiteness of N = 4 Yang-Mills. The oldest argument relies on the fact that in supersymmetric theories the superconformal anomalies lie in a supermultiplet. Another argument uses a generalization of the well known non-renormalization theories of N = 1 supersymmetry to extended supersymmetry. The third argument uses light-cone techniques. We will discuss the first two arguments in turn and refer the reader to reference (9) for a discussion of the light cone argument.

THE ANOMALIES ARGUMENT[10]

This argument for the finiteness of N = 4 Yang-Mills theory makes two assumptions which are as follows:

(a) In the quantum theory N = 1 supersymmetry and SU(4) internal symmetry are preserved;

(b) When viewed as an N = 1 theory the anomalies must belong to a chiral multiplet of the form

$$\left[\vartheta_\mu{}^\mu, \partial^\mu j_\mu^{(5)}, \gamma^\mu j_{\mu\alpha}, C, D\right] \tag{2.1}$$

In the above multiplet $\vartheta_{\mu\nu}$ is the energy momentum tensor, $j_\mu^{(5)}$ the chiral current (the R current), $j_{\mu\alpha}$ is the supercurrent and C and D are objects of dimension three for which no interpretation is known. This anomaly multiplet represents the breaking, due to quantum effects, of dilation, chiral and S supersymmetry invariance.

The finiteness of N = 4 Yang-Mills is given by the following simple argument. Assumption (a) implies that the SU(4) currents are preserved. However, N = 4 Yang-Mills is only SU(4) invariant not U(4) invariant and consequently all the nine chiral currents of the theory are preserved. The fact that N = 4 Yang-Mills does not possess an additional U(1) invariance follows from the CPT self-conjugate nature of the N = 4 Yang-Mills multiplet and the relation between the generator, B of this extra U(1) and the supersymmetry charge $Q_{\alpha i}$, namely

$$\left[Q_{\alpha i}, B\right] = \frac{(N-4)}{4}(\gamma_5)_\alpha^\beta Q_{\beta i} \tag{2.2}$$

which is zero for N = 4. Consequently, in any N = 1 decomposition of the theory the R current will be preserved, i.e. $\partial^\mu j_\mu^{(5)} = 0$.

Assumption (b) then implies that $\vartheta_\mu^{\ \mu} = 0$, which in turn implies that $\beta(g) = 0$.

The finiteness of the theory in a background field calculation or possibly with a particular gauge choice then follows from $\beta(g) = 0$.

The above argument can also be formulated in terms of N = 2 supersymmetry. In this case the assumptions are that N = 2 supersymmetry and O(4) internal symmetry be preserved in the quantum theory and that the N = 1 anomaly multiplet of equation (2.1) is replaced by a corresponding N = 2 supermultiplet. We refer the reader to reference (10) for the details of this argument.

The assumptions made above have been found to be correct in all models for which the results are known.[11] However, these results are restricted to N = 1 models, and for the above argument to become complete it is necessary to establish the validity of these assumptions for N = 4 Yang-Mills. We will return to this point later. It is interesting to note, however, that the above argument would apply to any theory in which the chiral current is preserved and its divergence sits in the same multiplet at the trace of the energy momentum tensor.

THE NON-RENORMALIZATION ARGUMENT

This argument is based on the use of extended super-Feynman rules and the associated non-renormalization theorems. Let us consider N = 4 Yang-Mills in terms of N = 2 superfields; it then consists of N = 2 Yang-Mills containing physical component fields $(C, D, \lambda_{\alpha i}, A_\mu)$ represented[12] by a superfield potential $A_{\alpha i}(\chi_\mu, \vartheta_{\beta j})$ and N = 2 matter (the hypermultiplet) containing physical component fields $(A_i, B_i, \chi_{\alpha i})$ and represented by the superfield[13] $\varphi_i(\chi^\mu, \vartheta_{\beta j}, z)$; i = 1, 2. When carrying out an N = 2 super-Feynman rule calculation in the background field formalism any contribution to the effective action must be

(a) an integral over all superspace

(b) a gauge invariant local function of $A_{\alpha i}$ and φ_i.

In other words any contribution must be of the form

$$\int d^4x d^8\vartheta L(A_{\alpha i}, \varphi_i, D_{\alpha j}\varphi_i, \cdots).$$

where L is a gauge invariant function. Consequently, any infinite counter-term for the superfield $A_{\alpha i}$ must be of the form

$$\int d^4x d^8\vartheta\, A_{\alpha i}\, D_{\gamma k} \cdots D_{\delta e}\, A_{\beta j} \tag{2.3}$$

However since $A_{\alpha i}$ and $D_{\gamma k}$ have dimension $(\text{mass})^{1/2}$ and the integration measure has dimension zero such a term is impossible. A similar argument applies to counter-terms involving the hypermultiplet φ_i which has dimension $(\text{mass})^1$ or indeed any infinite counterterms. In this context it is interesting to note that in the case of N = 1, the argument fails as the integration measure has dimension $(\text{mass})^2$ so allowing the gauge invariant term

$$\int d^4x d^4\vartheta\, A_{\alpha i}\, D^2 A^{\alpha i} \tag{2.4}$$

Examining the background field method for extended supersymmetry, however, requires us to refine the above argument. In extended supersymmetry one finds, having fixed the gauge and found the relevant ghosts, that the ghosts themselves have a gauge invariance. This new gauge invariance requires new ghosts which in turn have a gauge invariance. This process in fact goes on indefinitely requiring an infinite number of ghosts. Fortunately, since these new ghosts only couple to background fields this infinity of ghosts only affects the one loop contribution to the effective action.

We may conclude from the above argument that N = 4 Yang-Mills is finite above one loop. The finiteness of N = 4 Yang-Mills then follows since it is known by explicit calculation to be finite at one loop.[8]

An argument for the finiteness of N = 4 Yang Mills along similar lines to that given above, but involving N = 4 superfields can be found in reference 14. This reference also contains a discussion of non-renormalization theorems and the background field for extended supersymmetry.

Like the anomalies argument the above N = 2 argument has an assumption which in this case is the existence of an N = 2 superfield background field formalism for N = 4 Yang-Mills.

The validity of this assumption was more subject to doubt than the above argument suggests for two reasons. First, the field is subject to constraints which must be solved in terms of an unconstrained superfield which can be used to formulate super-Feynman rules. The solution of this constraint in the abelian case[15] is given in terms of a dimension $(\text{mass})^{-2}$ superfield U^{ij} and more recently it has been shown[16,17] that this solution can be systematically iterated to provide the non-abelian solution. The second problem concerns the occurrence of the extra bosonic coordinate, z in the φ^i formulation[13] of the hypermultiplet. This coordinate is not intergrated over in the action and represents an off-shell central charge.[18] It is not known how to construct Feynman rules for such superfields. Fortunately, an alternative description of N = 2 matter which does not involve off-shell central charges has been found:[16,19] It involves superfields S, L^{ij} and L^{ijkl} which are of dimension $(\text{mass})^1$ and subject to constraints which can be solved.

With the solution of these two problems an N = 2 background superfield formalism for N = 4 Yang-Mills when expressed in terms of N = 2 superfields is guaranteed and the non-renormalization argument for the finiteness of N = 4 Yang-Mills given above can be regarded as complete.

The N = 2 background field formalism for N = 4 Yang-Mills theory can also be used to justify the assumptions made in the anomalies argument. It is evident from the formalism that the N = 4 Yang-Mills theory, when viewed as an N = 2 theory has a manifest supersymmetry and U(2) internal symmetry. These symmetries will also be manifest at the quantum level provided the theory is regulated in a supersymmetric, and U(2) invariant fashion. Such a regulator is provided above one loop by higher covariant derivatives. In fact, it has also been explicitly checked in the case of N = 4 Yang-Mills that the higher derivatives do not introduce any new infinities at one loop.[20] The supermultiplet of anomalies must be a dimension $(\mathrm{mass})^3$ gauge invariant function of the N = 2 superfields. A preliminary analysis confirms that this multiplet is indeed of the correct form in that it contains at least one of the divergences of the U(2) chiral currents as well as the trace of the energy momentum tensor. An application of the anomalies argument to these facts gives the desired finiteness. Further details of this will be given elsewhere.

The reader may be struck by the fact that the anomaly argument as given just above does not seem to use any of the properties of N = 4 Yang-Mills other than one loop finiteness and the N = 2 superspace formulation of the theory. This line of reasoning leads to the results of the next section.

3. A CLASS OF FINITE N = 2 SUPERSYMMETRIC THEORIES

The results presented in this section have been found in collaboration with P. Howe and K. Stelle.[21] The anomalies and non-renormalization arguments for the finiteness of N = 4 Yang-Mills presented in the previous section, have been given in a form which used an N = 2 decomposition of N = 4 Yang-Mills. Examination of the arguments shows that they apply not only to N = 4 Yang-Mills, but to an arbitrary N = 2 theory consisting of N = 2 Yang-Mills coupled to N = 2 matter (the hypermultiplet), provided there exists a background superfield formalism for an arbitrary N = 2 theory. One can, however, demonstrate that it is possible to extend the new formulation of the hypermultiplet to an arbitrary representation provided there are an even number of hypermultiplets. The hypermultiplet is then represented by the complex superfields S, L^{ij}, L^{ijkl} which again are of dimension $(\mathrm{mass})^1$. The reader is referred to reference (21) for more details. This being the case we may conclude that an arbitrary N = 2 theory consisting of N = 2 Yang-Mills and N = 2 matter will be finite above one loop, although infinities can occur at one loop as indeed they do in the case of N = 2 Yang-Mills by itself.

The coupling of N = 2 matter to N = 2 Yang-Mills in the absence of U(1) factors is entirely determined by the gauge coupling constant. As a consequence, (see the next section), the necessary and sufficient condition for finiteness is that the Callan-Symanzik beta function vanish at one loop. The one loop beta function for a supersymmetric theory consisting of n_i Wess-Zumino multiplets (N = 1 chiral superfields) in the representation R_i of a gauge group G and one super Yang-Mills multiplet is given by[22]

$$\beta(g) = \frac{g^3}{16\pi^2}\left[\sum_i n_i T(R_i) - 3C_2(G)\right] \tag{3.1}$$

An arbitrary N = 2 theory is composed from N = 2 Yang-Mills, which consists of: N = 1 Yang-Mills and one Wess-Zumino multiplet in the adjoint representation, as well as N = 2 matter which consists of two Wess-Zumino multiplets, one in the representation R_i and the other in the complex conjugated representation $\overline{R}_i$. Consequently the β function for such a theory with m_i hypermultiplets in the representation $R_i(\overline{R}_i)$ is given by

$$\beta(g) = \frac{2g^3}{16\pi^2}\left[\sum_i m_i T(R_i) - C_2(G)\right] \tag{3.2}$$

The equation $\beta(g) = 0$ has many solutions. One example is given by taking G to be SU(N) and R_i to be the fundamental representation; (in this case $C_2(N) = N$ and $T(R_i) = \frac{1}{2}$; the resulting theory is finite provided

$$M = 2N \tag{3.3}$$

Another example is provided by taking G to be SU(5) where finiteness occurs if the N = 2 matter belongs to four $5(\overline{5})$ dimensional and two $10(\overline{10})$ dimensional representations. Yet another example is if G is SO(10) and the N = 2 matter consists of four 16 dimensional representations.

Let us summarize the results of this section; an arbitrary N = 2 theory possessing spins less than or equal to one is finite above one loop and as consequence the one loop result for the β function is exact. Those theories for which the condition $\sum_i m_i T(R_1) = C_2(G)$ holds will be finite to all orders.

One can play similar games in other supersymmetric theories. One example is N = 4 conformal supergravity, the fields for which are contained in a superfield $W(\chi,\vartheta)$, which satisfies the condition $D_{\alpha i} W = 0$. The action for this theory is given by

$$\int d^4x d^8\vartheta\, W^2 + h.c. \tag{3.4}$$

Since this integral is not over the whole of superspace we may expect this theory to be finite above one loop. Finiteness at one loop has been found by explicit calculation.[23]

None of the above arguments apply to supergravity due to the dimensional character of the gravitational coupling constant.

4. AN EXPLICIT PROOF OF TWO LOOP FINITENESS OF AN ARBITRARY N = 2 SUPERSYMMETRIC THEORY

The calculation given in this section has been performed in collaboration with P. Howe.[24] We will show by explicit calculation that a arbitrary N = 2 theory with spins less than or equal to one is finite *at* two loops despite in some cases having infinities at one loop. This serves as check on the general arguments given in the previous section.

The method we will adopt is to write the N = 2 theory in terms of N = 1 superfields and use the N = 1 super-Feynman rules. The N = 2 Yang-Mills multiplet breaks up into N = 1 Yang-Mills represented by the superfield V and an N = 1 Wess-Zumino multiplet represented by the chiral superfield φ^i in the adjoint representation. The N = 2 matter (the hypermultiplet) breaks into two Wess-Zumino multiplets represented by a chiral superfield X^a and in the representation R_i and a chiral superfield Y_b in the representation $\overline{R}_i$. The action for the arbitrary N = 2 theory being considered can then be expressed in the form

$$A = \int d^4x d^4\vartheta \Big[Tr\left(e^{-gV}\overline{\varphi} e^{gV}\varphi\right)$$

$$+ Tr\left\{ -\frac{1}{2}V \quad V + gV/16\left\{D^\alpha V, \overline{D}^2 D_\alpha V\right\} + \cdots \right\}$$

$$+ \overline{X}_a \left(e^{gV}\right)^a{}_b X^b + \overline{Y}^a \left(e^{-gV}\right)^b{}_a Y_b \Big]$$

$$+ \left[\int d^4x d^2\vartheta X^a Y_b (\varphi^i T_i)^b{}_a + h.c.\right]$$

$$+ \textit{ghost and gauge-fixing terms} \tag{4.1}$$

Examining the above action one may ask why the φXY interaction term must have coupling strength, g. The reason is that the N = 2 supersymmetry relates φ to V and X to Y and so this term must have the same strength as the gauge coupling $g\overline{X}XV$.

The fields φ^i, X^a, Y_b and the coupling constant g renormalize multiplicatively, namely

$$\varphi^i \to z_\varphi^{1/2}\varphi^i, \quad X \to z_x^{1/2}X, \quad Y \to z_y^{1/2}Y$$

$$g \to z_g g \tag{4.2}$$

The interaction between the two N = 2 multiplets is of the form

$$g \int d^2\vartheta d^4x\, \varphi XY \tag{4.3}$$

This term, by the N = 1 renormalization theorems[1,2] is not renormalized. As a result we may conclude that

$$z_g z_x^{1/2} z_y^{1/2} z_\varphi^{1/2} = 1. \tag{4.4}$$

The strategy we will adopt is to calculate the $\overline{X}X$, $\overline{Y}Y$ and $\overline{\varphi}\varphi$ propagators at two loops and show that the two loop contribution to z_x, z_y and z_φ, are zero. It then follows from equation (4.4) that z_g has no two loop contributions and consequently $\beta(G)$ has no two loop contribution.

It is interesting to note that in connection with the above argument that in a background field calculation $z_g z_V = 1$. This fact, taken in conjunction with the relations $z_V = z_\varphi$ and $z_x = z_y$ which are given by the O(2) invariance, implies that $z_\varphi z_g = 1$ and hence that $z_x = z_y = 1$.

It is also instructive to calculate the one loop beta function in this manner. The relevant graphs for the $\overline{X}X$ and $\overline{\varphi}\varphi$ propagator are given in Figures (4.1) and (4.2) respectively. The result for the $\overline{X}X$ and propagator is easily found to be

$$C_2(R)\overline{X}_a\{\,[1]-[1]\,\}\, ln\ \Lambda X_a = 0 \tag{4.5}$$

While the result for the $\overline{\varphi}\varphi$ propagator is given by

$$\overline{\varphi}_i\left\{\left[\sum_j m_j T(R_j)\right]-\left[C_2(G)\right]\right\}\varphi_i = \overline{\varphi}_i\varphi_i\left[\sum_k m_r T(R_k) - C_2(G)\right] \tag{4.6}$$

The VV propagator gives a similar result. We recover the result of the previous section namely the theory is finite at one loop provided

$$\sum_i m_i T(R_i) = C_2(G)\,. \tag{4.7}$$

We note that in this case the one loop contributions to the propagators vanish.

The graphs contributing to the two loop $\overline{\varphi}\varphi$ propagator are given in Figure (4.3). In the figure shaded vertices represent one loop insertions which are renormalized and the vertex with an "x" represents the one loop counter terms. Regularization is achieved by first integrating over the ϑ's and then using dimensional regularization to define the momentum integrals. This is none other than the dimensional reduction regularization scheme.[25] One finds after lengthy calculation that the $\overline{\varphi}\varphi$ propagator is finite *at* two loops so z_φ receives no two loop contribution. Similarly one finds that the $\overline{X}X$ and $\overline{Y}Y$ propagator is finite. From the previous argument we may conclude that $\beta(g)$ has no two loop contribution.

The above is a very brief sketch of this calculation and the words "finite at two loops" are understood in a special sense; the interested reader is referred to reference (24) for more details.

It is important to realize that the arguments given in this article are based on the perturbation theory and it is not ruled out that the beta function could acquire non-perturbative contributions. Hopefully these theories in which the beta function is known to all orders of perturbation theories will prove useful in testing general ideas concerning field theories. Another interesting avenue to explore is whether these theories can be solved exactly using superconformal Ward identities.

5. FINITENESS AND EXPLICIT SYMMETRY BREAKING IN N = 4 YANG-MILLS

The work reported in this section has been performed in collaboration with A. Parkes.[26,27] We wish to address the following question: can we add symmetry breaking terms to N = 4 Yang-Mills and still maintain the finiteness properties of the theory. We will consider adding all possible gauge invariant terms of

Figure 4.1. The one loop $\bar{\chi}\chi$ propagator.

Figure 4.2. The one loop $\bar{\varphi}\varphi$ propagator.

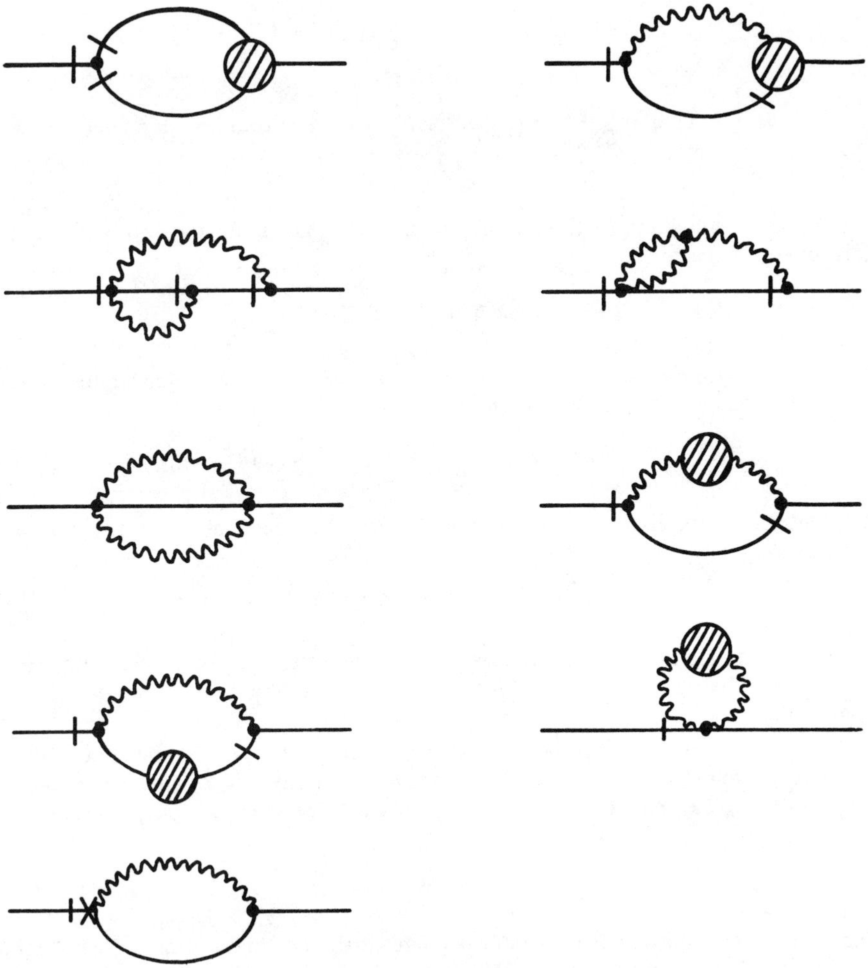

Figure 4.3. The two loop $\bar{\varphi}\varphi$ propagator.

dimension three or less and derive the conditions under which finiteness holds.

The preliminary tool for this analysis is to express N = 4 Yang-Mills in terms of N = 1 superfields and use the N = 1 super-Feynman rules. The N = 4 Yang-Mills theory is composed of one N = 1 Yang-Mills represented by a general real superfield V which contains component fields A_μ, λ_α, D as well as gauge degrees of freedom, and three N = 1 Wess-Zumino multiplets represented by chiral superfields φ_i ; i=1,2,3 which contain component fields $A_j + iB_j, \chi_{\alpha j}, F_j + iG_j$. The N = 4 Yang-Mills action can then be expressed in the form[28,29,30]

$$A = Tr \int d^4x d^4\vartheta \left\{ \bar{\varphi}^i \varphi_i - \frac{1}{2} V \quad V + g\left[\bar{\varphi}^i, V\right] \varphi_i \right.$$

$$\left. + \frac{1}{16} g V \left\{ D^\alpha V, \bar{D}^2 D_\alpha V \right\} + \frac{1}{2} g^2 \left[[\bar{\varphi}^i, V], V \right] \varphi_i + \cdots \right\} \tag{5.1}$$

$$+ Tr \left\{ \int d^4x d^2\vartheta \frac{ig}{3!} e^{ijk} \varphi_i [\varphi_j, \varphi_k] + h.c. \right\} + \textit{ghost and gauge fixing terms}$$

The Feynman rules and further details of this form of the action can be found in References (29) and (30).

ADDITION OF AN N = 1 SUPERSYMMETRIC MASS TERM

Let us begin by adding a term which preserves one of the four supersymmetries, namely an N = 1 mass term which is of the form

$$Tr \left(\int d^2\vartheta m^{ij} \varphi_i \varphi_j + h.c \right) = m^{ij} \, Tr F_i A_j - G_i B_j \, \bar{\chi}_i \chi_j \tag{5.2}$$

Elimination of the auxiliary fields F_i and G_i leads to a component expression of the general form

$$m^2(A^2 + B^2) + m\bar{\chi}\chi + gmA(A^2 + B^2) \tag{5.3}$$

as well as the original quartic, m independent terms in the action. It is important to note that the N = 1 mass term involves terms which are cubic in the spin zero fields.

It is relatively simple to prove that the addition of this term does not destroy the finiteness of the theory. The divergence of any N = 1 super-Feynman graph with E_{ch} external chiral lines and $C < \varphi\, \varphi >$ (or $< \bar{\varphi}\, \bar{\varphi} >$) propagators is given by[31]

$$D = 2 - C - E_{ch} \tag{5.4}$$

The $< \varphi\varphi >$ propagator, although not present in the original action is introduced by the addition of the mass term of Equation (5.2). The propagators of the theory are now of the form

$$< \varphi\, \varphi> = \frac{M}{k^2(k^2 + m^2}D^2\delta^4(\vartheta - \vartheta')$$

$$< \varphi\, \varphi^* > = \frac{1}{k^2 + m^2}\delta^4(\vartheta - \vartheta')$$

$$< V\, V >= - \frac{1}{k^2}\delta^4(\vartheta - \vartheta') \qquad (5.5)$$

Examining equation (5.4) tells is that if $E_{ch} > 2$ the graphs are finite and if $E_{ch} = 2$, then $D = - C$ and the graphs are finite if $C \neq 0$. The case E_{ch} with no external V lines is forbidden on grounds of gauge invariance. Graphs with E_{ch} with external vector lines must be gauge invariant, that is of the form $VD^4V \cdots V$, and since we have removed a D^4 factor from the inside of the graph the degree of divergence will be -C. There do exist graphs with $E_{ch} = 1$ and external vector lines, but their divergence is suppressed by the same argument.

Using these facts we recover the well known statement that N = 1 supersymmetric theories only have logarithmic divergences. We can also conclude that the m dependent contribution to any graph will be superficially finite. This follows once we identify the possible sources of m dependence. One possibility source is the occurrence of m in a $< \varphi > \varphi >$ propagator, but in this case $C \neq 0$ and by the above argument the graphs are superficially finite. The only other possibility is when m occurs in the $< \varphi\, \overline{\varphi} >$ propagator. However, as

$$< \varphi\, \varphi^* > = \frac{1}{k^2 + m^2}\delta_{12} = \left\{ \frac{1}{k^2} + \frac{m^2}{(k^2)^2} + \cdots \right\}\delta_{12}$$

any resulting power of m^2 occur with a $\frac{1}{k^2}$ factor and any m dependence in the graphs which were at most logarithmically divergent will be superficially finite.

We are now in a familiar situation studied first by Weinberg[32] in the context of Q.C.D. Arguing order by order, let us assume that all n loop graphs are rendered finite by using mass independent renormalization constants. Then, as m dependent terms are superficially finite the infinites in any n + 1 loop graph can be subtracted using m independent renormalization constants. We note that subintegrations of the n + 1 loop graphs are finite by assumption. From the above comments it should be clear that the one loop infinities are independent of m and so by induction the theory as whole can be renormalized using m independent renormalization constants.

Now the term we have added, $Tr \int d^4x d^2\vartheta \varphi^2 + h.c.$ is an integral over a subspace of superspace and so it is not renormalized.[1,2] Consequently adding this term introduces no new infinities and the infinities of the theory, being independent of m, are just those of N = 4 Yang-Mills which is finite. Hence adding an N = 1 supersymmetric mass term maintains the finiteness of the theory.

ADDITION OF TERMS WHICH HAVE NO SUPERSYMMETRY

To evaluate the effects on the ultraviolet structure of adding terms which have no supersymmetry at all we will use the spurion technique[33] in the context of the N = 1 formulation of the N = 4 Yang-Mills of Equation (5.1). The advantage of this technique is that it enables us to continue using the N = 1 super-Feynman rules. This is achieved by using a directional superfield to write the term that explicitly breaks supersymmetry. Consider adding to the action quantities such as

$$\int d^4x d^4\vartheta\, UX(\varphi_i, V, D_\alpha\varphi_i, \cdots) \tag{5.6}$$

$$\int d^4x d^2\vartheta\, NT(\varphi_i, V, D_\alpha\varphi_i, \cdots) + \text{h.c.} \tag{5.7}$$

where supersymmetry is broken by the choice

$$U = \vartheta^2\bar{\vartheta}^2 \cdot constant \text{ and } N = \vartheta^2 \cdot constant \tag{5.8}$$

The superfields X and T are taken to be gauge invariant functions of the chiral superfields φ^i and general superfield V and covariant derivatives of these fields. One could add terms which explicitly break the gauge group,[34] but we will consider this possibility. It is easy to see that the dimension of X is 2 - dim U and that of T is 3 - dim N.

Let us first consider adding terms of the type given in Equation (5.6) to the action of Equation (5.1). All possible terms induced in the effective action by the insertion of one spurion field will be of the form

$$\Lambda^{P_1} \int d^4x d^4\vartheta\, Uf_1(\varphi_i, V, D_\alpha\varphi_i, \cdots) \tag{5.9}$$

where Λ is the cut-off or some similar quantity representing the divergences of the expression. Using the fact that the dimensions of f_1 is greater than or equal to two (as $Tr\,\varphi_i = 0$) we find that

$$P_1 \le -2 + dim\ X \tag{5.10}$$

The terms induced by the insertion of many spurions will be of the form

$$\Lambda^{P_{n+1}} \int d^4x d^4\vartheta\, U \left[D^2\, \bar{D}^2 U\right]^n f_2(\varphi_i, V, D_\alpha\varphi_i, \cdots) \tag{5.11}$$

where the covariant derivatives on U are required since $UD_\alpha\bar{D}^2U = 0$. The degree of divergence, P_{n+1} of this expression is easily found to be

$$P_{n+1} \le -2 - 4n + (n+1) dim\ X \tag{5.12}$$

We note that for insertions with dim $X \le 4$ we have that $P_{n+1} \le p_n$ for all n => 1.

We now consider the addition of a term of the type given in Equation (5.7). The insertion of one N will lead to an expression of the form

$$\Lambda^{r_1} \int d^4x d^4\vartheta\, Ng_1(\varphi_i, V, D_\alpha\varphi_i, \cdots) \tag{5.13}$$

in the effective action. The degree of divergence is found to be

$$r_i \leq -3 + dim\ T \tag{5.14}$$

A super-graph with s N insertions and t $\overline{N}$ insertions will give rise to a term in the effective action of the form

$$\Lambda^{r_{s,t}} \int d^4 \times d^4\, \vartheta N\overline{N}(D^2N)^{s-1}(\overline{D}^2\overline{N})^{t-1} g_2(\varphi_i, V, D_\alpha\varphi_1, \cdots) \tag{5.15}$$

The degree of divergence is given by

$$r_{s,t} \leq -4(t+s) + (t+s)\, dim\ T + 2 \tag{5.16}$$

This power counting procedure is equivalent to the result obtained by counting the number of momentum factors occurring in the N = 1 super Feynman graphs and then calculating the overall degree of divergence. We will not consider insertions in vacuum graphs, which can only lead to infinite constant terms in the effective action.

The entries in the table below summarize the maximum possible degree of divergence that results from various choices of X and T.

dim X or	Insertion				
dim T	U	$U(D^2D^{-2}U)^M$	N	$\overline{N}N$	$N\overline{N}(D^2N)^{s-1}(D^2\overline{N})^{t-1}$
2	0	-2n	-1	-2	$2-2(t+s)$
3	1	-n+1	0	0	$2-(t+s)$
4	2	+2	1	2	2

These results use the fact that X or T must have dimension not less than two, since $Tr\,\varphi_i = 0$ and V must occur in a gauge invariant manner. Using the table above, we will now consider the various insertions in increasing order of hardness; some of the results will leave the theory finite while others will give rise to possible logarithmic infinities.

As a first example, let us consider adding a term of the form

$$Tr\left\{\int d^4x d^2\vartheta N^{ij}\varphi_i\varphi_j + h.c.\right\} = 2Tr \int d^4x \eta^{ij}(A_iA_j - B_iB_j) \tag{5.17}$$

where $N^{ij} = \vartheta^2\eta^{ij}$. The super-Feynman rules are the same as for N = 4 Yang-Mills except for the extra vertex corresponding to the term in Equation (5.17).

From the previous discussion and the fact that N^{ij} has dimension $(mass)^1$ it is clear that there are no possible counterterms, and by the same argument as was used for the N = 1 supersymmetric mass the addition of the term of Equation (5.17) does not introduce any infinities.

Let us now consider adding the only other dimension two object, namely a mass term of the form $A^2 + B^2$. In spurion form this requires us to add a term of the form

$$Tr \int d^4x d^4\vartheta e^{-igV} \overline{\varphi}^i e^{gV} \varphi_s U_i^s = Tr\left\{\mu_i^s \left[A^i A_s + B^i B_s\right]\right\} \tag{5.18}$$

where $U_i^s = \vartheta^2 \overline{\vartheta}^2 \mu_i^s$. This term leads to the new vertices given in Figure 5.1 where the spurion line $U_i j$ is represented by a dashed line.

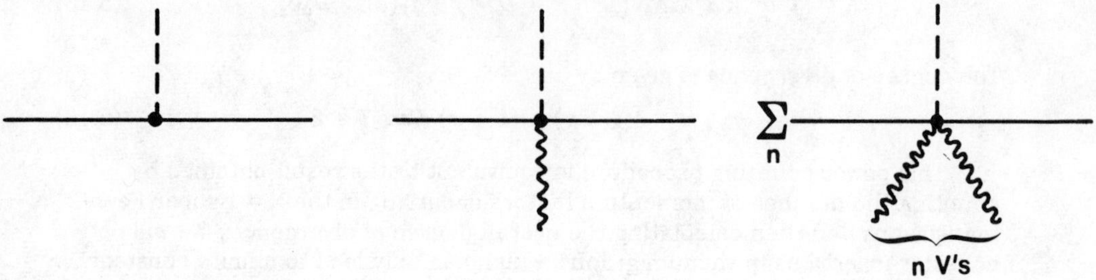

Figure 5.1. The new vertices introduced by a mass term of the form $A^2 + B^2$.

The infinities generated in the effective action can only occur in the form

$$Tr \int d^4\vartheta U \overline{\varphi} \varphi \tag{5.19}$$

An infinity of the form $Tr\left[d^4\vartheta u \overline{\varphi}^2 + h.c.\right]$ is ruled out by the counting rule

$$E_\varphi - E_{\overline{\varphi}} = 3(V_\varphi - V_{\overline{\varphi}})$$

where E_φ $(E_{\overline{\varphi}})$ are the number of external chiral (antichiral) lines and $V_\varphi (V_{\overline{\varphi}})$ are the number of cubic chiral (anti-chiral) vertices in the contributing graph.

The addition of terms of dimension three proceeds along similar lines; the possible counterterms being determined by the dimension of the spurion field as well as the relevant counting rules. In what follows we will suppress indices, traces and integrations and only give the generic features of the terms being added and the counterterms they produce.

The addition of a mass term for the fermions $\chi_{\alpha i}$ belonging to the N = 1 chiral superfields φ_i is achieved by adding an insertion of the form

$$Tr\left\{\int d^4\vartheta M^{ij} D^\alpha \varphi_i D_\alpha \varphi_j + \text{h.c.}\right\} = \text{Tr}\left\{\text{m}^{ij} \chi_i^\alpha \chi_{\alpha j} + \text{h.c.}\right\} \tag{5.20}$$

where $M^{ij} = \vartheta^2 \overline{\vartheta}^2 m^{ij}$. This gives rise to the following possible logarithmically infinite counterterms

$$A^2 + B^2,\ A^2 - B^2,\ A(A^2 + B^2),\ FA - GB \tag{5.21}$$

The last infinity, FA - GB can be rewritten by eliminating the auxiliary fields

F and G in the effective action. The result of this elimination neglecting terms of the order $\hbar^2$ is to give a term of the form $A(A^2 + B^2)$ This can be added to the term of this form which was already present. We list in the table below the insertions along with the possible infinities they give rise to. Any auxiliary fields have been eliminated in favor of the physical fields A_i and B_i.

	Infinity Produced					
Insertion	A^2-B^2	A^2+B^2	$\bar{\chi}\chi$	$\bar{\lambda}\lambda$	$A(A^2+B^2)$	A^3-3AB^2
A^2-B^2						
A^2+B^2		√				
$\bar{\chi}\chi$	√	√				√
$\bar{\lambda}\lambda$	√	√				√
$A(A^2+B^2)$	√	√			√	
A^3-3AB^2	√	√				√

The above analysis only takes into account one of the four possible residual supersymmetries of N = 4 Yang - Mills as well as the semi-simple nature of the gauge group. It is possible that the additional properties of this remarkable theory could lead to cancellations among the possible divergences given in the table above. To examine this possibility we will calculate the infinities at the one loop level.

The $A^2 + B^2$ insertion of Equation (5.18) gives rise to an infinity at one loop which is given by the one loop graphs of Figure 5.2.

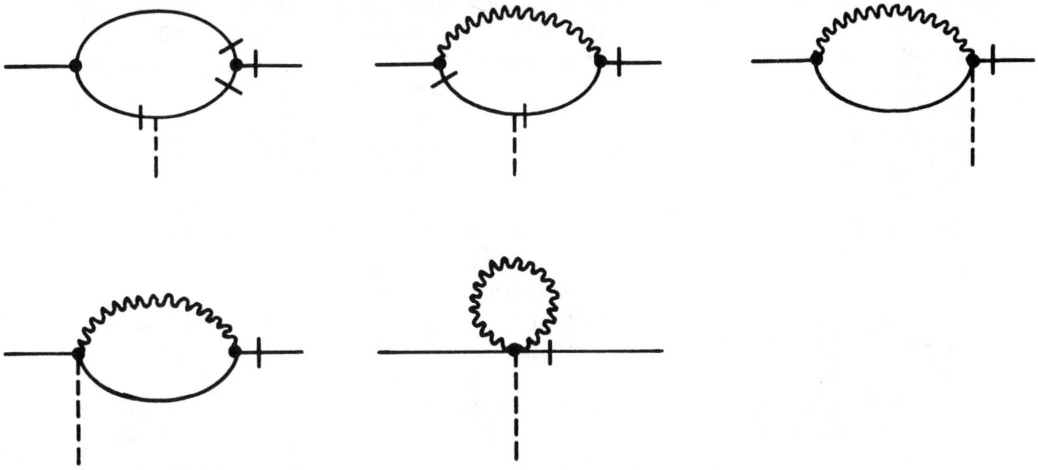

Figure 5.1. The one loop graphs which might give an infinity due to the $A^2 + B^2$ insertion.

Evaluating these graphs leads to the following one loop counterterm

$$g^2\overline{\varphi}^s\left\{\left[U_s^i - \delta_s^i U_k^k\right] + \left[u_s^i\right] + \left[-U_s^i\right] + \left[-U_s^i\right] + \left[0\right]\right\} = -g^2\overline{\varphi}^i\varphi_i U_k^k \quad (5.22)$$

There will be no one loop infinity provided

$$\sum_k \mu_k^k = 0 \quad (5.23)$$

Taking diagonal mass terms this allows for two free parameters.

The results of calculating all the one loop infinities, in a similar manner, are as follows. The $\bar{\chi}\chi$ insertion of Equation (5.20) leads to all of the possible infinities that can arise. Examination of the table shows that the only way to cancel the $A(A^2 + B^2)$ infinity is also to add to the theory a term of the form

$$Tr\left\{A_i\left(\left[A^i, A^k\right] + \left[B^j, B^k\right]\right)\right\}\left(i\eta^i_{jk}\right).$$

Calculation shows that this infinity only vanishes provided the coefficient η^i_{jk} of the cubic insertion and the mass m^{ij} for $\bar{\chi}\chi$ are related by the relation

$$\eta^k_{ij} = 2gm^{kr}\varepsilon_{rij} \quad (5.24)$$

Adopting this relation we find that the $A^2 - B^2$ infinities these two insertions produce cancel.

The $A^2 + B^2$ infinity can be cancelled by adding an $A^2 + B^2$ insertion with an appropriate coefficient. This coefficient turns out to have precisely the value necessary and sufficient for the supertrace of the mass squared to vanish. In fact, these relations necessary for the finiteness are none other than the insertions required for an N = 1 supersymmetric mass (see Eq. 5.3) which we have already established to maintain finiteness to all orders.

Further adding the $\bar{\lambda}\lambda$ insertion we find that it also gives all possible infinities of which the $A^3 - 3AB^2$ infinity can only be cancelled by the additional insertion of a $A^3 - 3AB^2$ term. The necessary condition for finiteness is that the coefficient of the cubic term is determined by the mass of the spinor λ. Again the $A^2 - B^2$ infinities then cancel automatically and we can readjust the coefficient of the $A^2 + B^2$ insertion so as to remove the additional $A^2 + B^2$ infinities. This adjustment is just so as to re-establish the condition that the super-trace of the squares of the masses be zero. It follows from the way the infinities cancel that the order in which one adds the insertions does not affect the final result.

In effect we find that the $A^2 - B^2$ insertion leads to no new infinities. Each fermion mass term requires for finiteness the addition of a unique term cubic in the fields A_i and B_i. Once this is arranged the $A^2 - B^2$ infinities which are produced cancel automatically. The $A^2 + B^2$ infinities cancel provided the coefficient of the $A^2 + B^2$ insertion is adjusted so that the super-trace of the squares of the masses is zero.

If we take the masses to be diagonal, simple counting shows that we are allowed *nine* free mass parameters and can still have a finite theory. These nine parameters contain 3 masses of the form $A^2 - B^2$, and 3 N = 1 masses for the chiral multiplet φ_i which in components are of the form

$$A^2 + B^2 + \bar{\chi}\chi + A(A^2 + B^2) .$$

These six parameters maintain the finiteness of the theory to all orders. Another mass parameter is accounted for by the mass term for the spinor λ with its corresponding cubic term $A^3 - 3AB^2$. Due to the O(4) relation between this term and the χ mass term we may be confident that this result, which has only been proven at one loop extends to all orders. Finally we have the two parameters corresponding to the freedom in Equation (5.23). These last two parameters have only been shown to maintain finiteness at one loop.

The above is a somewhat schematic account and the reader is referred to references (26) and (27) for more details.

The results of the last two sections raise an obvious question; can we add explicit symmetry breaking terms to the N = 2 class of finite theories? It is apparent from the above discussion that one can add mass terms of the form $A^2 - B^2$ and N = 1 supersymmetric mass terms without ruining finiteness. What other terms one can add is under investigation.

Clearly it is of interest to investigate the types of gauge symmetry breaking patterns that can emerge from the addition of the terms considered in this section. It would be interesting to examine if breaking patterns similar to that found in References (35) and (36) can emerge.

ACKNOWLEDGEMENT

I wish to thank P. Howe, A. Parkes and K. Stelle for valuable discussions.

NOTE ADDED

After the lecture presented in this article had been given, we received a paper from M.A. Namazie, A. Salam and J. Strathdee[37] who, using lightcone techniques have found similar results to those presented in section 5.

REFERENCES

1. J. Wess and B. Zumino. Phys. Lett. **49B**, 52(1974). J. Iliopoulos and B. Zumino, Nucl. Phys. **B76**, 1310 (1974) S. Ferrara, J. Iliopoulos and B. Zumino, Nucl. Phys. **B77**, 41 (1974).
2. S. Ferrara and O. Piguet. Nucl. Phys. **B93**, 261 (1975). D. Capper and G. Leibbrandt. Nucl. Phys. **B85**, 492 (1975) K. Fujikawa and W. Lang. Nucl. Phys. **B88**, 61 (1975). R. Delbourgo Nuovo Cimento 25A, 646 (1975). M.T. Grisaru, W. Siegel and M. Rocek. Nucl. Phys. **B159**, 42 (1979).
3. For a review of irreducible representation: see D.Z. Freedman in: Recent Developments in Gravitation, Cargese 1978, eds. M. Levy and S. Deser, Plenum Press; P.C. West Supergravity '81 edited by S. Ferrara and J.G. Taylor, Cambridge University press, also S. Ferrara the same volume.

4. S. Deser and B. Zumino. Phys. Lett. **62B**, 335 (1976).
5. A. Salam and J. Strathdee. Phys. Lett. **51B**, 33, (1974). P. Fayet. Nucl. Phys. **B113**, 135 (1976).
6. P. Fayet. Nucl. Phys. **B113**, 135 (1976).
7. F. Gliozzi, D. Olive and J. Scherk. Nucl. Phys. **B122**, 253 (1977). L. Brink, J. Schwarz and J. Scherk. Nucl. Phys. **B121**, 77 (1977).
8. S. Ferrara and B. Zumino. Nucl. Phys. **B79**, 413 (1974). D.R.T. Jones. Phys. Lett. **72B**, 199 (1977). E. Poggio and H. Pendleton. Phys. Lett. **72B**, 200 (1977). O. Tarasov, A. Vladimirov. A. Yu. Phys. Lett. **93B**, 429 (1980). M.T. Grisaru, M. Rocek and W. Siegel. Phys. Rev. Lett. **45**, 1063 (1980). W.E. Caswell and C. Zanon. Nucl. Phys. **B182**, 125 (1981).
9. S. Mandelstam, Berkeley preprint UCB-PTH-82115, (1982); also see contribution in this volume. L. Brink, O. Lindgren and B. Nilsson Texas preprint UTTG-1-82.
10. M. Sohnius and P. West. Phys. Lett. **100B**. 245 (1981). S. Ferrara and B. Zumino, unpublished.
11. T. Clark, O. Piguet and K. Sibold. Ann. of Phys. **109**, 418 (1977). Nucl. Phys. **B143**, 445 (1978); Nucl. Phys **B159** (1979).
12. R. Grimm, M. Sohnius and J. Wess. Nucl. Phys. **B133**, 275 (1978).
13. M. Sohnius. Nucl. Phys. **B165**. 483 (1980).
14. M. Grisaru and W. Siegel. Nucl. Phys. **B201**, 292 (1982).
15. L. Mezincesu - JINR Report P2-12572 (1979).
16. P. Howe, K. Stelle and P. Townsend, in preparation.
17. J. Koller "Unconstrained Prepotentials in Extended Superspace" Caltech preprint - CALT-68-981.
18. M. Sohnius, K. Stelle and P. West. Phys. Lett. **92B**, 123 (1980); Nucl. Phys. **B173**, 127 (1980).
19. P. Howe, K. Stelle and P. Townsend "The relaxed hypermultiplet: an unconstrained N = 2 superfield theory". Nucl. Phys. B. to be published; P. Howe, K. Stelle, P. Townsend, in preparation.
20. K. Stelle Proceedings of the Parish High Energy Conference 1982, Imperial preprint.
21. K. Stelle, P. Howe and P. West "A class of finite four-dimensional supersymmetric field theories" Santa Barbara preprint NSF-ITP-83-09 (Phys. Lett. B to be published).
22. S. Ferrara and B. Zumino. Nucl. Phys. **B79**, 413 (1974).
23. E.S. Fradkin and A.A. Tseytlin. Phys. Lett. **110B**, 117 (1981).
24. P. Howe and P. West in preparation.
25 W. Siegel. Phys. Lett. **84B**, 193 (1979).
26. A. Parkes and P. West "N = 1 supersymmetric Mass Terms in the N = 4 supersymmetric Yang-Mills Theory" King's preprint (Phys. Lett. B. to be published).

27. A. Parkes and P. West "Finiteness and explicit supersymmetry breaking in the N = 4 supersymmetric Yang-Mills Theory"; King's preprint.

28. P. Fayet. Nucl. Phys. **B149**, 137 (1979).

29. M. Grisaru, M. Rocek and W. Siegel. Nucl. Phys. **B159**, 429 (1979). Nucl. Phys. **B183**, 141 (1981).

30. W.E. Caswell and D. Zanon. Nucl. Phys. **B182**, 125 (1981).

31. S. Ferrara and O. Piguet. Nucl. Phys. **B93**, 261 (1975). D. Capper and G. Leibbrandt. Nucl. Phys. **B85**. 492 (1975). K. Fujikawa and W. Lang. Nucl. Phys. **B88**, 61 (1975).

32. S. Weinberg. Phys. Rev. **D8**, 3497 (1973).

33. L. Girardello and M. T. Grisaru. Nucl Phys. **B194**, 55 (1982).

34. Some terms which explicitly break gauge invariance are considered in J.G. Taylor "Soft breaking in N = 4 Yang-Mills Theory" King's preprint.

35. D. Olive and P. West "The N = 4 supersymmetric E_8 gauge theory and coset space dimensional reduction" Imperial preprint (Nucl. Phys. B to be published).

36. G. Chapline and R. Slansky; Nucl. Phys. B to be published.

37. M.A. Namazie, A. Salam and J. Strathdee "Finiteness of Broken N = 4 super-Yang-Mills theory" Trieste preprint.

No.26	High-Energy Physics and Nuclear Structure - 1975 (Santa Fe and Los Alamos)	75-26411	0-88318-125-8
No.27	Topics in Statistical Mechanics and Biophysics: A Memorial to Julius L. Jackson (Wayne State University, 1975)	75-36309	0-88318-126-6
No.28	Physics and Our World: A Symposium in Honor of Victor F. Weisskopf (M.I.T., 1974)	76-7207	0-88318-127-4
No.29	Magnetism and Magnetic Materials - 1975 (21st Annual Conference, Philadelphia)	76-10931	0-88318-128-2
No.30	Particle Searches and Discoveries - 1976 (Vanderbilt Conference)	76-19949	0-88318-129-0
No.31	Structure and Excitations of Amorphous Solids (Williamsburg, VA., 1976)	76-22279	0-88318-130-4
No.32	Materials Technology - 1976 (APS New York Meeting)	76-27967	0-88318-131-2
No.33	Meson-Nuclear Physics - 1976 (Carnegie-Mellon Conference)	76-26811	0-88318-132-0
No.34	Magnetism and Magnetic Materials - 1976 (Joint MMM-Intermag Conference, Pittsburgh)	76-47106	0-88318-133-9
No.35	High Energy Physics with Polarized Beams and Targets (Argonne, 1976)	76-50181	0-88318-134-7
No.36	Momentum Wave Functions - 1976 (Indiana University)	77-82145	0-88318-135-5
No.37	Weak Interaction Physics - 1977 (Indiana University)	77-83344	0-88318-136-3
No.38	Workshop on New Directions in Mossbauer Spectroscopy (Argonne, 1977)	77-90635	0-88318-137-1
No.39	Physics Careers, Employment and Education (Penn State, 1977)	77-94053	0-88318-138-X
No.40	Electrical Transport and Optical Properties of Inhomogeneous Media (Ohio State University, 1977)	78-54319	0-88318-139-8
No.41	Nucleon-Nucleon Interactions - 1977 (Vancouver)	78-54249	0-88318-140-1
No.42	Higher Energy Polarized Proton Beams (Ann Arbor, 1977)	78-55682	0-88318-141-X
No.43	Particles and Fields - 1977 (APS/DPF, Argonne)	78-55683	0-88318-142-8
No.44	Future Trends in Superconductive Electronics (Charlottesville, 1978)	77-9240	0-88318-143-6
No.45	New Results in High Energy Physics - 1978 (Vanderbilt Conference)	78-67196	0-88318-144-4
No.46	Topics in Nonlinear Dynamics (La Jolla Institute)	78-057870	0-88318-145-2
No.47	Clustering Aspects of Nuclear Structure and Nuclear Reactions (Winnepeg, 1978)	78-64942	0-88318-146-0
No.48	Current Trends in the Theory of Fields (Tallahassee, 1978)	78-72948	0-88318-147-9
No.49	Cosmic Rays and Particle Physics - 1978 (Bartol Conference)	79-50489	0-88318-148-7

No.50	Laser-Solid Interactions and Laser Processing - 1978 (Boston)	79-51564	0-88318-149-5
No.51	High Energy Physics with Polarized Beams and Polarized Targets (Argonne, 1978)	79-64565	0-88318-150-9
No.52	Long-Distance Neutrino Detection - 1978 (C.L. Cowan Memorial Symposium)	79-52078	0-88318-151-7
No.53	Modulated Structures - 1979 (Kailua Kona, Hawaii)	79-53846	0-88318-152-5
No.54	Meson-Nuclear Physics - 1979 (Houston)	79-53978	0-88318-153-3
No.55	Quantum Chromodynamics (La Jolla, 1978)	79-54969	0-88318-154-1
No.56	Particle Acceleration Mechanisms in Astrophysics (La Jolla, 1979)	79-55844	0-88318-155-X
No. 57	Nonlinear Dynamics and the Beam-Beam Interaction (Brookhaven, 1979)	79-57341	0-88318-156-8
No. 58	Inhomogeneous Superconductors - 1979 (Berkeley Springs, W.V.)	79-57620	0-88318-157-6
No. 59	Particles and Fields - 1979 (APS/DPF Montreal)	80-66631	0-88318-158-4
No. 60	History of the ZGS (Argonne, 1979)	80-67694	0-88318-159-2
No. 61	Aspects of the Kinetics and Dynamics of Surface Reactions (La Jolla Institute, 1979)	80-68004	0-88318-160-6
No. 62	High Energy e^+e^- Interactions (Vanderbilt , 1980)	80-53377	0-88318-161-4
No. 63	Supernovae Spectra (La Jolla, 1980)	80-70019	0-88318-162-2
No. 64	Laboratory EXAFS Facilities - 1980 (Univ. of Washington)	80-70579	0-88318-163-0
No. 65	Optics in Four Dimensions - 1980 (ICO, Ensenada)	80-70771	0-88318-164-9
No. 66	Physics in the Automotive Industry - 1980 (APS/AAPT Topical Conference)	80-70987	0-88318-165-7
No. 67	Experimental Meson Spectroscopy - 1980 (Sixth International Conference , Brookhaven)	80-71123	0-88318-166-5
No. 68	High Energy Physics - 1980 (XX International Conference, Madison)	81-65032	0-88318-167-3
No. 69	Polarization Phenomena in Nuclear Physics - 1980 (Fifth International Symposium, Santa Fe)	81-65107	0-88318-168-1
No. 70	Chemistry and Physics of Coal Utilization - 1980 (APS, Morgantown)	81-65106	0-88318-169-X
No. 71	Group Theory and its Applications in Physics - 1980 (Latin American School of Physics, Mexico City)	81-66132	0-88318-170-3
No. 72	Weak Interactions as a Probe of Unification (Virginia Polytechnic Institute - 1980)	81-67184	0-88318-171-1
No. 73	Tetrahedrally Bonded Amorphous Semiconductors (Carefree, Arizona, 1981)	81-67419	0-88318-172-X
No.74	Perturbative Quantum Chromodynamics (Tallahassee, 1981)	81-70372	0-88318-173-8

No. 75	Low Energy X-ray Diagnostics-1981 (Monterey)	81-69841	0-88318-174-6
No. 76	Nonlinear Properties of Internal Waves (La Jolla Institute, 1981)	81-71062	0-88318-175-4
No. 77	Gamma Ray Transients and Related Astrophysical Phenomena (La Jolla Institute, 1981)	81-71543	0-88318-176-2
No. 78	Shock Waves in Condensed Matter - 1981 (Menlo Park)	82-70014	0-88318-177-0
No. 79	Pion Production and Absorption in Nuclei - 1981 (Indiana University Cyclotron Facility)	82-70678	0-88318-178-9
No. 80	Polarized Proton Ion Sources (Ann Arbor, 1981)	82-71025	0-88318-179-7
No. 81	Particles and Fields - 1981: Testing the Standard Model (APS/DPF, Santa Cruz)	82-71156	0-88318-180-0
No. 82	Interpretation of Climate and Photochemical Models, Ozone and Temperature Measurements (La Jolla Institute, 1981)	82-071345	0-88318-181-9
No. 83	The Galactic Center (Cal. Inst. of Tech., 1982)	82-071635	0-88318-182-7
No. 84	Physics in the Steel Industry (APS.AISI, Lehigh University, 1981)	82-072033	0-88318-183-5
No. 85	Proton-Antiproton Collider Physics - 1981 (Madison, Wisconsin)	82-072141	0-88318-184-3
No. 86	Momentum Wave Functions - 1982 (Adelaide, Australia)	82-072375	0-88318-185-1
No. 87	Physics of High Energy Particle Accelerators (Fermilab Summer School, 1981)	82-072421	0-88318-186-X
No. 88	Mathematical Methods in Hydrodynamics and Integrability in Dynamical Systems (La Jolla Institute, 1981)	82-072462	0-88318-187-8
No. 89	Neutron Scattering - 1981 (Argonne National Laboratory)	82-073094	0-88318-188-6
No. 90	Laser Techniques for Extreme Ultraviolt Spectroscopy (Boulder, 1982)	82-073205	0-88318-189-4
No. 91	Laser Acceleration of Particles (Los Alamos, 1982)	82-073361	0-88318-190-8
No. 92	The State of Particle Accelerators and High Energy Physics(Fermilab, 1981)	82-073861	0-88318-191-6
No. 93	Novel Results in Particle Physics (Vanderbilt, 1982)	82-73954	0-88318-192-4
No. 94	X-Ray and Atomic Inner-Shell Physics-1982 (International Conference, U. of Oregon)	82-74075	0-88318-193-2
No. 95	High Energy Spin Physics - 1982 (Brookhaven National Laboratory)	83-70154	0-88318-194-0
No. 96	Science Underground (Los Alamos, 1982)	83-70377	0-88318-195-9

No. 97	The Interaction Between Medium Energy Nucleons in Nuclei-1982 (Indiana University)	83-70649	0-88318-196-7
No. 98	Particles and Fields - 1982 (APS/DPF University of Maryland)	83-70807	0-88318-197-5
No. 99	Neutrino Mass and Gauge Structure of Weak Interactions (Telemark, 1982)	83-71072	0-88318-198-3
No. 100	Excimer Lasers - 1983 (OSA, Lake Tahoe, Nevada)	83-71437	0-88318-199-1
No. 101	Positron-Electron Pairs in Astrophysics (Goddard Space Flight Center, 1983)	83-71926	0-88318-200-9
No. 102	Intense Medium Energy Sources of Strangeness (UC-Santa Cruz, 1983)	83-72261	0-88318-201-7
No. 103	Quantum Fluids and Solids - 1983 (Sanibel Island, Florida)	83-72440	0-88318-202-5
No. 104	Physics,Technology and the Nuclear Arms Race (APS Baltimore-1983)	83-72533	0-88318-203-3
No. 105	Physics of High Energy Particle Accelerators (SLAC Summer School, 1982)	83-72986	0-88318-304-8
No. 106	Predictability of Fluid Motions (La Jolla Institute, 1983)	83-73641	0-88318-305-6
No. 107	Physics and Chemistry of Porous Media (Schlumberger-Doll Research, 1983)	83-73640	0-88318-306-4
No. 108	The Time Projection Chamber (TRIUMF, Vancouver, 1983)	83-83445	0-88318-307-2
No. 109	Random Walks and Their Applications in the Physical and Biological Sciences (NBS/La Jolla Institute, 1982)	84-70208	0-88318-308-0
No. 110	Hadron Substructure in Nuclear Physics (Indiana University, 1983)	84-70165	0-88318-309-9
No. 111	Production and Neutralization of Negative Ions and Beams (3rd Int'l Symposium, Brookhaven, 1983)	84-70379	0-88318-310-2
No. 112	Particles and Fields-1983 (APS/DPF, Blacksburg, VA)	84-70378	0-88318-311-0
No. 113	Experimental Meson Spectroscopy - 1983 (Seventh International Conference, Brookhaven)	84-70910	0-88318-312-9
No. 114	Low Energy Tests of Conservation Laws in Particle Physics (Blacksburg, VA, 1983)	84-71157	0-88318-313-7
No. 115	High Energy Transients in Astrophysics (Santa Cruz, CA, 1983)	84-71205	0-88318-314-5
No. 116	Problems in Unification and Supergravity (La Jolla Institute, 1983)	84-71246	0-88318-315-3
No. 117	Polarized Proton Ion Sources (TRIUMF, Vancouver, 1983)	84-71235	0-88318-316-1